Disaster Psychiatry In Haiti:
Training Haitian medical professionals

KENT RAVENSCROFT, MD

ISBN	978-1-105-52850-7
Copyright	lulu.com 2012 (Standard Copyright License)
Edition	first
Publisher	lulu.com 2011
Published	February 19, 2012
Language	English
Pages	176
Binding	Perfect-bound Paperback
Interior Ink	Black & white
Dimensions (inches)	6.0 wide x 9.0 tall

Kent Ravenscroft, age 20, Leogane Haiti 1961

Dr. Kent Ravenscroft, age 70, Petionville Tent City 2010

Dedication

I wish to thank Sidney Mintz, my Yale anthropology professor and Haiti specialist; Carole Richardot who suggested I study Haiti and whose family was my family away from home in Haiti; Maya Deren, a brilliant Haiti writer and dance ethnographer who trusted me enough to introduce me to Odette and Milo Mennesson-Rigaud, foremost Haitian Vodun experts who arranged my field sites and gave me access to the Vodun inner sanctum; my friend and informant, Bonheur Calixte, top voodoo drummer in Masson; Dr. Mario Maj, President of the World Psychiatric Association, who was crucial in referring me to the International Medical Corps (IMC) the superb medical disaster relief organization taking me to Haiti as a medical volunteer; and to their great trio of psychiatrists, Drs. Lynne Jones, Nick Rose, and Peter Hughes, who selected, trained, and launched me out to the Petit Goave mobile medical clinics; Tessier, my great translator, Nathalie my psychosocial nurse; Stephanie our fearless IMC Petit Goave Director; all the Haitian IMC doctors and nurses I trained in disaster psychiatry (and who trained me in the realities of disaster medicine and patient care), and finally, the Haitian people who gave me so much as an anthropologist, and now allowed me to care for them as I worked along side and taught their Haitian doctors and nurses.

Preface

As any effective community or consultation-liaison psychiatrist knows, it is best to understand the history and culture of the people you are treating. To have that intimate perspective allows more accurate work. Whether you are doing disaster psychiatry in the inner city or a foreign country, it is extremely helpful to know whom you are working with as you enter their world to help them with their tragedies. As an example, here is a brief selective history of Haiti, something you will see come alive as we go out to the countryside to do our relief work, our brand of disaster psychiatry:

Haiti is again at the crossroads--and in the cross hairs. How history repeats itself. Drugs and corruption, hurricanes and mudslides, and now an earthquake followed by cholera torment her people.

She was once a lush tropical paradise called the Pearl of the Antilles, with beautiful forested mountains and rich alluvial plains-- the crown jewel among France's prosperous colonies.

Her rich and hungry plantations devoured vast numbers of slaves, many dying on slave ships before reaching their harsh new world. To subdue arrivals, French slave masters broke and scattered their families and tribes, rendering them totally dependent on their overseers.

But the slave masters made one fatal mistake. They allowed the slaves to hold religious ceremonies in the dead of night. From shared African roots, ancient and powerful Rada gods, cool and wise, appeared during ceremonies, biding their time. Through spirit possession, an adapted form of cultural multiple personality from Africa, these gods took over the consciousness of the downtrodden slaves, one at a time during ceremonies, speaking boldly of ancestor worship, community and hope. Slowly, the new world religion of Voodoo, or Vodun, was born and spread among the plantations through out Haiti. Led by powerful priests called Houngans, Voodoo became a healing and unifying force for downtrodden slaves during their darkest hours.

Even so, life on the plantations became more unbearable near the end of the 18th century. Suddenly, hot impatient new Petro gods began possessing slaves, demanding blood and revenge. Out of this Voodoo hotbed, a volcanic eruption shook Haiti, as these bloodthirsty gods and their angry Houngans ignited rebellion across the land. After thirteen years of guerilla war, aided by malaria and yellow fever, Haiti's slaves stunned Europe with a military victory over 20,000 troops sent by Napoleon -- becoming the only slave colony to win freedom, making Haiti the second country and first black nation to gain independence in the Western Hemisphere.

From this glorious moment in 1804, Haiti has experienced a tragic decline. Her succession of corrupt and macabre governments has allowed her people to plunge into abject malaria-ridden poverty. With virtually no roads or phones, no trains or power, Haiti's hearty but illiterate peasants barely survive. Stripping the mountains of trees for charcoal, they now suffer dangerous erosion, flooding and mudslides. Thousands are buried alive during tropical storms and hurricanes. Though the plains are still fertile, overpopulation and excessive land division have left the peasants eking out their existence. With few doctors and the highest infant mortality rate in the western world, life is so difficult that hundreds of boat people try to escape to the United States each year--abandoning their beloved island of blood and bougainvilleas. Launching from vast stretches of totally unpoliced shoreline, many join their slave ancestors in watery graves, while others are turned back by the US Coast Guard. The most impoverished nation in the Western Hemisphere despite a port-of-call for glittering cruise ships, Haiti has no effective police or judiciary. Starved and helpless, she lies there, ripe for the narco-plucking and prone to disaster.

With the January 2010 earthquake, a disaster of unimaginable proportions struck Haiti to the quick, followed soon by a debilitating Cholera epidemic. Immediately after the quake, relief groups began pouring in. Amid death, destruction, and dislocation, tent cities began to spring up, including the 50,000 person Petionville Tent City run by Sean Penn, perched high above Port-au-Prince, on the Petionville Country Club golf course. With this brief history and anthropology in mind, I prepared myself for lay ahead.

.
.

Introduction

Much is written about disaster psychiatry[1], all essential reading for anyone volunteering to do this kind of work. Much less is written about what it is actually like to do the work on the ground under these dire circumstances. How do you actually apply this important body of knowledge to traumatized patients and families? And do it when you yourself are anxious, tired, and pushed to the limit? How do you manage your own health and mental health while working in the trenches alongside equally stressed colleagues, when all of you are struggling with flooded clinics, minimal equipment and short supplies? How do you keep your head screwed on straight when the work itself is traumatizing, and the circumstances crazy making? How do you cope as you question your clinical skills and competence, your judgment calls, and your ethical standards at every turn? The circumstances keep changing the rules of the ball game as you struggle to do your best under constantly fluctuating circumstances, despite the best efforts of whatever NGO (non-governmental organization) you're working with.

This book is about the Haiti earthquake, a disaster of unimaginable proportions. It is about the experience of one psychiatrist, a volunteer with the International Medical Corps (IMC), who arrived soon after the January 12, 2010, 7.1 earthquake. While having volunteer doctors do some direct service, the IMC's major volunteer mission was to train Haitian physicians and nurses to do better on-going clinical work themselves. The IMC follows the old adage: Give a person a fish and he eats for a day; teach a person to fish and he eats for a lifetime.

Every volunteer physician or nurse, every psychiatrist, brings his or her own 'baggage', own discipline and experience, own strengths and weaknesses, resulting in a range of unique volunteer experiences. Yet there is a common thread to all of this, weaving a story worth telling for future volunteers to consider. By walking in another volunteer's moccasins during Haiti psychiatric disaster

---

[1] Disaster Psychiatry: Readiness, Evaluation, and Treatment [Paperback] Frederick J. Stoddard (Author, Editor), Anand Pandya (Editor), Craig L. Katz (Editor) April 15, 2011 | ISBN-10: 0873182170 | ISBN-13: 978-0873182171

work, you can better prepare yourself for what lies ahead as you embark on your own volunteer effort. Hopefully this will lessen your culture shock, improve your clinical skills, and deepen your satisfaction as you do this demanding work.

While surgeons and medical doctors have certain defenses allowing them to do their arduous trauma work, these same (necessary) self-protections make them variably immune and unaware of certain other things; psychiatrists, to do their intimate emotional work with trauma victims, have to let down their own guard and become more open to their patients' and their own inner experience, potentially putting them at greater risk in disaster situations. But this openness also gives psychiatrists unique perspectives. These insights can be useful to surgeons and internists, as well as psychiatrists, and to anyone interested in disaster work in general.

Embedded in the ongoing narrative of this book are most of the principles and practices of disaster psychiatry. Though the specific approaches taken represent one particular practitioner's way of doing things, guided by IMC principles, general principles and application techniques emerge around a wide range of gripping cases.

At heart, this is the remarkable story of a seasoned older psychiatrist who once lived in Haiti as a young Yale anthropologist, returning now to spend a grueling month on the front lines of the Haitian relief effort—an experience confronting him with unexpected medical and personal challenges, and exposing him to buried corners of his own mind.

By offering a candid first-hand description of his time in Haiti--making us feel we are there in the tent and in clinic with him--he provides a memorable basis for understanding and applying the principles of disaster psychiatry.

His journey begins in France.

Tuesday, February 16, 2010: Foie Gras and Fate

I was sitting there quite alone at our crowded table when the call came. We were in the Dordogne enjoying foie gras and truffles. I was trying to forget the earthquake that had ravaged my beloved Haiti. I didn't want to ruin the **Les Liaisons Delicieuses** trip my wife had worked so hard to create. But some 50 years earlier, as a Yale undergraduate anthropologist, I had lived in Haiti with a voodoo priest and his family just outside Leogane—now the epicenter of the quake. I had kept in touch with my friends there, recently receiving first hand reports of massive death and destruction. Sad and guilty, I feared I would never be asked to join the medical relief effort despite my attempts, wondering in passing if it were because I was now 70.

My French cell phone vibrated in my pocket, jolting me back to reality. Embarrassed, I turned and cupped my hand over the phone. "Who is it?"

"Drs. Lynne Jones and Peter Hughes from Haiti with the International Medical Corps. Do you have Skype?"

"Yes, down in my room."

"Could we call you back in a few minutes? What's your Skype name?"

Twenty minutes later I reappeared at the table, after a pointed interview about my psychiatric background and perspectives on disaster psychiatry. They had actually quizzed me on a hypothetical case of a dazed incoherent Haitian woman found naked wandering the streets of Port-au-Prince.

"Why are you looking so ashen?" my wife asked. 'What just happened?"

"They want me to come to Haiti. They're starting up mental health teams near Leogane for the mobile emergency medical clinics."

"I thought that's just what you wanted?" Rod Drake, my closest psychiatric friend in Washington added.

"I never thought it would actually happen." Then I recalled my surgeon brother-in-law, Mike Ribaudo's warning, "Be careful what you wish for." Well here it was, the die was cast. Apparently, the Washington Psychiatric Society and World Psychiatric Association through its President Mario Maj had sent my volunteer resume on to the International Medical Corps.

There was a hitch, though. After being processed to go, I received an email from the International Medical Corps deployment officer saying I had been 'put on extended hold' because of overstaffing. Hurt and miffed, I emailed back an impassioned rejoinder, "Take me in the next few days or I'll go elsewhere!" I didn't want to miss my chance. Luckily I copied Lynne Jones.

Lynne fired back immediately, "Hold on Kent, this isn't coming from me. Let me see what I can do." Within one hour I was off hold and back on track--my first taste of what a brilliant bulldog Lynne could be. She's a British Child Psychiatrist, a great clinician-teacher, and crack administrator. I soon learned nobody messes with her. *Thank god somebody like her is down there in Haiti,* I thought.

Thursday, March 4, 2010: Paris to Port-au-Prince: Time Warp

During my check-in for departure from Paris, Air France charged me $300 overweight for my medical supplies and equipment. I even reminded them Haiti had been their former colony but to no avail. Aboard the plane, seated on my left was a French *sapeur pompier*, with his large fire and rescue team behind him, ready to do relief combat and water purification. On my right was one of the top people in the World Health Organization who told me that within the first two weeks in Haiti he had to coordinate 240 humanitarian groups, medical and otherwise. He was returning after a break to deal with a much larger number, now some 900 strong, though some were pulling out as the most acute phase was ending.

Just before I left, my sister-in-law Polly, a neonatal nurse, talked to her colleague on the hospital ship Comfort, moored off the shore of Haiti, who told her, "We've been swamped with the worst cases I've ever seen, many dying, but many saved. I had one little girl with a horribly infected face, worms and maggots crawling out of her festering wounds. Polly, we've never seen ANYTHING like this. No war zone compares. But, you know, I've been moved to tears by the strength and spirit of the Haitians I've seen."

Just as I was packing up, my wife Patti looked me straight in the eye, "Do you really know what you're getting yourself into?"

My anxiety shot sky high, "Please, that's enough. I have too much to think about already."

For me, at least consciously, I was more worried about my so-so French, my rusty Creole, and my ability to help psychiatrically in the midst of such devastating physical tragedy. Finally, I said, "Going down to Haiti feels like the biggest final exam I've ever taken. I thought I was completely finished with things like this." I had been plagued by performance anxiety all my life, and had been glad to be done with it.

"I'm not worried about how you'll do," Patti said. "I'm worried about your health and survival." "What about your pulmonary embolus three years ago? And the Coumadin you're still on?"

"It'll be okay." I said. "My health's pretty good now. Stop being so anxious."

"Me? Look at your hands." We both stared down at them. They were trembling.

"I've always had that hereditary tremor, nothing new. You know that's why I didn't go into surgery."

"Come on. You're vibrating like a tuning fork."

"Okay, so I'm a little nervous, but I'm going. I have to. You know why. And I need your support."

"Just be very very careful."

In the air, my cabin conversation slowly gave way to personal reflection. I slipped on earphones, and leaned back. A subtle sadness filled me. Intimations of the loss of the Haiti I knew 50 years ago came into my mind. I had experienced Haiti back then in the youthful blush of naive enthusiasm. My brash denial of risk and danger allowed me to tour carefree all over the country. Now TV images of collapsed buildings with survivors being hauled from rubble floated through my mind. I had had glimpses of what I would be dealing with once I arrived. And I would be seeing it from my sobered vantage point now later in life. Even so, I felt a shudder ripple through me. Something deeper was troubling me. Just then, 'Haiti Cherie' came over my earphones, taking me back to glittering evenings dancing at the posh Petionville country club. I was able to push more depressing thoughts away with these pleasant memories.

That first summer in Haiti had been such a high, the stuff of dreams. I was in love and loved what I was doing. I chronicled everything, capturing it all in weekly love letters to my girlfriend Linda. I had met Maya Deren in Greenwich Village as a Yale junior. A brilliant cinematographer and writer, she had lived with a Voodoo priest, Isnard, during her Guggenheim grant studying Haitian Vodun dance, and authoring the fascinating book, **The Divine Horsemen: the living gods of Haiti.** My hope was to study Voodoo spirit possession and try to understand its behavioral content and psychodynamics. The project took on such major proportions I applied for Yale's Scholar of the House Program, freeing me from all class work save presenting my research monthly, and writing my Scholar of the House thesis. When they accepted me I was ecstatic. I was doing my pre-med courses on the side.

Through Maya Deren I met Odette and Milo Mennesson-Rigaud, foremost Haitian authorities on Vodun.

Fifty years ago I stepped off a long propeller-driven flight, weary and bedraggled, meeting them at the gate. That first evening they took me to a special Voodoo ceremony, a *Retirer en bas de l'eau* ceremony they wanted to see—involving possession by a dead ancestor called out of the Abyss for installation in a *Govi* jar, so that their spirit didn't remain submerged, trapped and wandering in limbo forever.

As my Creole progressed, to break me in they arranged an interim field site in Croix des Mission through their painter friend, Andre Pierre, now recognized as the first and foremost primitive painter of Haiti.

My room was on the 'yard' right next to the voodoo temple. Though I saw several ceremonies and got better at Creole, I didn't see the mosquito that gave me malaria, causing hallucinations and a 105 degree fever, cutting my stay short. I rode to the hospital in a gaily painted truck, a *camionette,* with swine tied up at my feet.

I had *Falciparum* malaria, commonly known as 'Black Water Fever', something you either live or die from, but are never plagued by recurrences. I was so excited about going to my permanent field site the next week that this didn't daunt me. I was raring to go.

Soon after, Odette took me out to Brache and Masson near Leogane (now the epicenter of the earthquake), and introduced me to Ternvil Calixte, a Voodoo priest (*houngan*), wife Joselia, and their children, where I would live for almost a year over two long summers.

Victor Calixte, his relative, was the most powerful *houngan* in Masson, and his son, Bonheur, was the top Voodoo drummer.

He became my friend and chief informant about all things voodoo. My surrogate Haitian family and their friends welcomed me with open arms, and a great deal of curiosity. Because I had a little money, I asked his wife, Joselia Calixte, to buy me precious goat meat at the market each week so that I didn't just have to eat rice and red beans every day. But when I made her cook it rare she was disgusted. "What are you, an animal? Eating raw meat! It's not healthy." I became their favorite form of entertainment, given all my odd ways.

During the nine months, spread over two long hot summers, I made wonderful friends and had countless adventures, attending Voodoo ceremonies and conducting extensive interviews. Since they knew I wanted to go to medical school, they granted me a degree and pressed me into medical service, asking me to treat everything from fevers and diarrhea to burns and a near-severed finger. It didn't take much to get me involved. My willingness endeared me to the peasants, who took me into the bosom of their families, despite my variable success rate.

Because my nickname back then was 'Sparky', they called me 'Ti Feu' or 'Little Fire' in Creole. I noticed when they introduced me to new peasants they often cracked a smile. Finally my friends clued me in. They were pronouncing my name ambiguously, sounding like 'Ti Fou', meaning a little 'crazy'. They found many things about me strange and funny, like my working in the mid-day heat, and even taking sunbaths, while they lounged or slept in the shade.

For R & R I would travel into Port-au-Prince to visit my friends the Richardots. I owed the inspiration for my research to Carole, their lovely daughter. She was the lab mate of my Yale roommates' girlfriend at Mount Holyoke. She suggested Haiti as a fascinating nearby culture for me to study for my Culture and Behavior major. Her father was the head of the United Nations Economic Commission to Haiti. I enjoyed my trips to their fabulous gingerbread house, and visits with all their sophisticated friends. They were always interested in what I was seeing, and in all my theories. It was amazing to walk out of exotic rural Haiti and into their house or the Petionville Club, hob-knobbing with the elite intelligentsia, the military and the NGOs bringing aid from other countries.

I loved writing Linda about all my experiences in the countryside, the voodoo ceremonies, the social dances (bals), the raucous domino games, the soccer matches, and the parade of medical problems brought to me. I wrote incessantly about the theories about voodoo spirit possession I was developing, probably driving her to distraction.

The only things I didn't tell her about were my constant bacillary dysentery that dropped my weight slowly from 190 to 150, and the ratty reddish blond beard and mustache I was growing--things that would surprise her. I even caught Dengue Fever at one point, feeling like all my bones and joints were breaking. But nothing stopped me. Toward the end of the first summer I met another powerful *Houngan*, Silvain, in a neighboring area.

Houngan Silvain standing next to possessed woman

His elaborate family ceremony worshipped many family deities (loa), who came to visit and give valuable advice through possessing family members. This ceremony was particularly noteworthy because it involved sacrifice of a bull.

Preparation for ritual sacrifice of the bull, done with great respect, as Bonheur drums

After such remarkable experiences, toward the end of the first summer I traveled extensively on my own, going out to Les Cayes near the tip of the southern peninsula and even up to Cap Haitien in the north, hiking past *Sans Souci*, Henri Christophe's grand palace constructed after he beat the French and won Haiti's independence, and on up to his Citadel, the massive mountaintop fortress he built, with huge canons trained on the distant bay lest the French invade again. I was on the move and loving every minute of it. Fascinating memories kept flowing through my mind.

Someone tapped my shoulder snapping me out of my reverie. "Would you like a snack?" the stewardess asked.

"No, I'm having a delicious time as it is," I said, slipping immediately back into my daydreams.

I was enjoying reliving the finale of my senior year. It was a blast. I had already pecked out my medical school applications on my typewriter down in Haiti. That fall I was accepted by Harvard. I gave presentations to Yale groups, published several articles, and made a trip back home to present my work at St. Louis Country Day School to inspire high school students. I worked steadily on my Scholar of the House thesis, 'Voodoo Spirit Possession: A Tentative Theoretical Analysis', which was well received in the spring. Things were going well.

Until I faced the gulf, the great divide—between the safety of Yale and the unknown of medical school. Rigor mortis began to set in. Unexpected demons lurked in the divides' abyss. My gap summer began to yawn like a forbidding chasm in front of me. What to do with the next four months before entering Harvard, and putting scalpel to my cadaver? Struggling with all this, my love for Linda began to die and a morbid drift set in. I finally broke up with her, putting me on a downhill skid. We were to have been in Boston together, but I became too anxious and estranged. Why did I drift away from such a wonderful person? I needed to have my head examined. One year later I did just that. It only took three years of psychotherapy and five of psychoanalysis to figure it out. Put in a nutshell, I was afraid of blood, guts and intimacy.

Faced with this vacuum, and completely at loose ends, I played Johnny-one-note, coat-tailing on my success. I decided to return to Haiti for a second long summer--even though my thesis was completed. It dumbfounds me to look back on this decision. What ever possessed me to return to Haiti after such a perfect first experience? Perhaps the word 'possessed' is the operative word here. Maybe one of the Voodoo gods got into my vacant head, giving me divine guidance at a decisive moment. As my daydreams took a macabre turn, I was starting to feel creepy thinking about all this.

Suddenly I was shaken out of my trance. From the corner of my eye I saw something black and hairy crawling onto my left shoulder. I screamed and sat bolt upright, wrenching myself away from it. I startled the hell out of the French Sapeur-Pompier next to me.

"Hey, buddy, you all right?"

"Sorry. Must have been a nightmare or something." I bumped the WHO guy next to me and he eyed me suspiciously. "I'll be okay. Just a bad memory." I settled back into my seat, not at all sure I wanted to close my eyes again. Haiti was rapidly coming closer. So was something else. As our jet approached Hispaniola, I found myself wishing we were in that old prop-driven plane I first took to Haiti long agao. I needed more time to get ready for all this. But I couldn't push certain memories away any longer.

My second summer in Haiti had been a rough one. I remember that fateful flight down, obsessing about breaking up with Linda--all my fault. I had given up a wonderful girlfriend because of my immaturity, and probably because I was more apprehensive about leaving the safety of college and embarking on medical school than I realized. If I thought medical school was a challenge, well, going back to Haiti now after an earthquake was many times more so, and I was beginning to think I might really be in for it. But the flight was long enough for something else to stir inside me. Old fault lines were beginning to shift. Apparently I had left a lot of unfinished business buried in Haiti. As different thoughts floated through my mind I felt myself tensing up. I began to take a more sober, even somber look at my life. I was starting to feel depressed, not just sad. Why?

During the process of writing my novel, **Body Sharing**, about Haiti and spirit possession, I had created a glossy, idealized cover story about my time in Haiti, something I began to believe myself, forgetting what actually happened to me down there. As Haiti approached, I was becoming unglued. How could I face the challenge of bringing emotional relief to the traumatized earthquake victims with my own tectonic plates sliding out from under me? My initial optimism was giving way to a sense of foreboding. Was I afraid I wasn't up to the challenge at my age, no longer capable of handling the demands and privations of hair-shirt front-line work? Caring for my peasant patients when I was young and knew nothing was a piece of cake. Now I knew too much. Or did I? Something was eating at me from inside. Patti's words, "Kent, do you know what you're getting yourself into?" rang in my ears, joining the tinnitus I already suffered because of anti-malarial medication I took in Haiti fifty years ago—something like hearing the steady low rumble of Haitian surf.

My mind began to drift, and a spate of unappetizing memories came floating back. The mysterious island of la Gonave loomed into view several times, sitting out there in Port-au-Prince Bay, held between the gapping jaws of the two peninsulas.

As my mood shifted, I began obsessing about why I had broken up. I quickly consoled myself with the thought, *Thank god I was so screwed up, because later I was blessed to find the true love of my life, my beloved Patti.*

Feeling bereft on the flight down that second summer, I happened to sit next to a vivacious college girl, Ruth Thurston, daughter of the new US Ambassador to Haiti. She told me about her weekly Port-au-Prince radio show. Things were suddenly looking up—until she blew me off. In retrospect I was a pest, trying to hang on to her in desperation. She wanted none of it. Once I arrived, I found myself having immense difficulty getting myself together enough to go out and face my field site. I shuddered at all its demands and difficulties. Using the flimsiest of excuses, I found myself avoiding getting started on my research work. My letters were to my parents and sister now, and I didn't have much to say. Instead of my project feeling exciting it felt onerous and intimidating.

As I sank deeper into my seat on Air France, headphones drowning out cabin noise, images of the mysterious Island of la Gonave percolated into my mind again. I remembered that fateful day long ago. Just after I arrived in Port-au-Prince that second summer I was invited to go sailing to Gonave on a spanking new yacht owned by a handsome, tan stockbroker who had cashed in his chips to buy his sleek new sloop. With him was a gorgeous young goddess, blond and bronzed, his mate for the adventure. We moored for the night next to a small island, the island of the giant iguanas. Because it was hot, I slept on deck, bothered occasionally by marauding mosquitoes. On the voyage back to Port-au-Prince, the stockbroker said, "We're looking for another deckhand for a spectacular trip we're planning, someone to help take watches and work closely with us. We plan to sail around Hispaniola, stopping at interesting ports in Haiti and the Dominican Republic. We'd love to have you join us." They both made it clear that they wanted to share more than their boat with me, the goddess soon shedding her clothes, sunbathing *au naturel* on the front deck, occasionally dropping back to the cabin for a cool drink. Her body language suggested a *ménage a trois* was in the offing. I was enchanted, and scared to death, tightening the drawstrings on my trunks. The name of their boat was **Mektub**, which they said meant 'Fate' in some Arabic language.

Since everything seemed so bleak, I found the stockbroker's offer tantalizing as I hiked back up to the Richardot's. A few nights later I couldn't sleep, tossing and turning, feverish and sweating profusely. I felt a sharp pain in my left chest. The pain became a vice-like grip tightening around my chest, morphing into a searing ache radiating up into my left shoulder and down my left arm. *My god, I'm having a heart attack. Or is it a panic attack over Mektub?* The pain increased. *No, this is real and I'm sick as hell.* When I tried to stand up to get help, everything went black for several seconds. I crumpled to the floor. Dawn was just breaking in the east. The Richardots took me to the hospital for an EKG, which showed paracarditis, but no heart attack. The Haitian doctor looked at me, "You have pleurodynia, mimics a heart attack. Probably the mosquito-borne virus Cocksackie B. You'll be weak for a while, but you'll be all right. Go home and convalesce."

With several days in bed, followed by lounging around the Richardot's, I had plenty of time to think, and lots of warm motherly attention. Probably too much. I began to feel like a malingerer, and even a phony. But I just didn't feel I could face going out to my field site. So I made my decision. I became the ex-stockbroker's latest sale. I called and accepted his offer, putting my fortunes with him and his goddess. I was just too depressed, lonely, and at loose ends to go out to the countryside on my own for four months. I was also aware the Richardots would be leaving soon, reassigned to another country. The world was changing and everything felt vastly different.

With the decision to join the **Mektub** crew I felt an immediate rush of excitement, a huge burden lifted from my shoulders. But almost immediately I had pangs of guilt. What about all the grant money I had been given for my research into Voodoo spirit possession? I just couldn't keep it. What should I do? Then it hit me. I decided to give it away to another anthropologist, partially solving my moral dilemma. Caroline Legermann, a bright Vassar undergraduate, also happened to be doing anthropological research on the southern peninsula that summer not far from me. I contacted our mutual mentor, Sydney Mintz, a famous Haiti expert, getting directions to her field site. Because she was blond and beautiful, she was unmistakable. The peasants knew exactly where she was, leading me right to her.

She was shocked to see me appear out of the blue. As she listened to my story, I could tell she thought I was sorry excuse for an anthropologist. But she was short on cash and accepted my money without comment. It was late in the day so I had to spend the night with her. There was nowhere for me to sleep but in the room with her. She knew exactly what her peasants would be thinking. And my fantasies were running wild. But in fact she offered me only a mat on the floor. Later I realized Sidney must have had the inside track, but under duress I was clueless. I spent a sleepless night for many reasons, slinking away early the next morning. The path back was long and torturous, taking much longer than I recalled. I imagined Ruth and Caroline teaming up on her radio show announcing to the world what a wimp I was. I began to obsess, feeling increasingly guilty and confused. Lost and ambivalent about everything in my life, I walked slowly back to the highway. What was I doing, and where was I going?

But character is destiny. On that long journey back to Port-au-Prince, in profound crisis, I finally had a change of heart. I just couldn't do it—sail off into uncharted waters, my moral compass spinning wildly. I decided to stick it out, go back to my field site near Leogane, and put my nose to the anthropology grindstone. And now I was virtually penniless.

With huge effort, I pulled myself together, grabbed all my stuff and headed out to Leogane. Finally we crossed the Momance River, and arrived at Brache, the little crossroads above my field site in Masson. Because I had sent word ahead, my friends and surrogate family had been expecting me for two weeks, and began worrying that I would never show. Word spread rapidly when I had arrived. I tried to smile but a cold sweat drenched my back and I developed a terrible stomach ache. *Tummy memories of all the dysentery,* I thought. I was there by dint of will, but my heart wasn't in it. Ternvil Calixte, my voodoo priest, came shuffling up the trail to greet me, led by his robust young wife, Joselia, who was periodically possessed by one of the most powerful voodoo gods, *Erzulie Gran Freda.*

She looked at me and grabbed my trunk, hefting it up on one strong shoulder, winking at me. All my friends pressed in, jabbering away in Creole. Suddenly I felt incredibly claustrophobic, and to my horror had massive diarrhea, drenching my khaki pants. I was humiliated beyond words. But no one seemed to notice. They crowded around me like a human shield as we moved en masse down the trail to Ternvil's temple and house. I finally said to hell with it, giving myself to the occasion. My spirit was willing even though my flesh was weak. No one bathed very often anyway. Getting there apparently scared the shit out of me. But I was back home with the people I loved.

At Ternvil's place, I made a hasty change, and carried on. It was a long hot grueling summer, and everyone knew I was somehow different. When they heard I had lost my dear Linda, they understood completely. I'm glad somebody did. As I put my nose to the grindstone, working on ethnography and family trees, both for peasant families and their god hierarchies, I slowly got back into the swing of things. I spent a lot of time studying the community of Louis Tore.

I was buoyed up by their kindness, love and support, taking heart as time went on. As word spread peasant patients began to trickle to my door. My side practice warmed my soul and confirmed my commitment to be a doctor, and not an anthropologist--or a nomad.

Penniless, I had to pass on goat meat the second summer, settling for lots of rice and red beans. Yum. I'm sure Joselia thought my taste for rare goat caused my diarrhea the first summer, but I still got bacillary dysentery anyway. Scientific proof. I slowly lost the same 40 pounds again. But I gained in moral stature, feeling substantially grounded by the end. Bonheur was a mainstay, bless his patient soul. We traded lessons in drumming and English. He learned to speak pretty well, but my drumming was atrocious. I also gave him a degree in psychotherapy for his steady support.

Two years ago, Karen Richman, an anthropologist friend of mine who worked in Masson sometime after I did gave me the upsetting news that Bonheur was dying of stomach cancer. He had come up to New York to see if an American hospital could save him, and was living with relatives in Brooklyn. I was deeply saddened and needed to see him. Patti and I made a pilgrimage to his house. With wide-eyed grandchildren in tow, we were ushered into a darkened back room. Bonheur was shriveled and hunched over a bedpan, spitting incessantly into it. He couldn't swallow any longer. "Bonheur, it's Ti Fou." His head snapped up, his eyes brightened, and then glistened with a rim of tears, a broad smile replacing his grimace.

"So you have come to see me, like old times." Then he caught sight of Patti behind me, and said, "Is that your Linda with you? My she is a beauty." He actually remembered Linda's name. I was impressed. He was still sharp. After explaining about Patti, we began reminiscing. Finally, I said, "Bonheur, I have a surprise for you. Take a look at this." Dramatically I pulled out a big picture. The family gathered around, the littlest ones wiggling in close to see.
There stood Bonheur in his powerful prime, handsome and robust, pith helmet cocked to one side, shirt off, muscles rippling, six-pack etched boldly, beating out a Voodoo rhythm on the Maman, the biggest of the Rada drums. Everyone gasped.

"Is that really you, Granpa? You're so big and strong!"

"And gorgeous!" one of the teenage girls added. Everyone clapped and the smile on Bonheur's face broadened.

"I remember that moment, 'EsSporky', his other name for me. "I think you still needed a few more drumming lessons." Everyone laughed.

I remembered vividly my time with Bonheur. As the last day at my field site rolled around, I felt eternally grateful. But the summer had been hard and draining, and I didn't want to let anyone know how glad I was to be leaving Haiti for good. I had visited the heart of darkness and was ready to escape back to civilization.

Apparently Ternvil had diagnosed me as needing a little divine guidance at the last moment, sensing my angst. He gave me a little private voodoo ceremony that last day, surprisingly without charging me a gourd (20 cents)—something unheard of for Ternvil. After a few cornmeal veve drawings and incantations, and the requisite shaking of his rattle, his *Ason*, he became possessed by a gurgling gravelly voiced voodoo god rising up out of the abyss to take over his mind and body.

This god stared at me with slit-like reptilian eyes and proceeded to tell me, "If you work very very hard at medical school, you will become a fine young doctor, even better than you are already." The god then fixed me with a piercing look and said something quite disturbing, given my urgency to get the hell out of Haiti.

"Ti Fou, if you become a really good doctor, someday you may have another chance to come back to Haiti and serve us. We have needed you now and we may well need you again. Never forget your friends here." Though this made me tense, I found myself tearing up with immense gratitude for all they had done to help me through that grueling summer of my discontent.

I thought to myself, *Maybe I'll come back some day, possibly as a tourist. But I don't think so. Right now I'm out of here, maybe for good. Two sultry summers are enough.*

I had forgotten how awful that second summer was—until these memories came flooding back in unbidden torrents on my Air France ride down. I began to realize returning to Haiti was not going to be so easy, not just because of the earthquake. The return of the repressed is always a deeply private, seismic event, causing fear and trembling. With these last thoughts, I began to wonder, *Just what kind of experience was this going to be—more like my first summer, or more like my second--more like a dream, or a nightmare? My future weighed in the balance.*

"Fasten your seatbelts. We will be landing in Port-au-Prince shortly," came the voice over the PA," I found myself tightening mine very tightly.

Later Thursday, March 4, 2010: Bumpy Landing

We were among the first commercial flights to land in Haiti, and I was one of the first child psychiatrists to arrive, after Lynne, and maybe someone with Partners-in-Health. It was a sobering thought. The airport was as chaotic as ever, entrepreneurial Haitians hawking baggage trolleys for $2 American, and several eager cabbies grabbing me and my stuff despite my protests, before I found my International Medical Corps driver. He grabbed one of my bags, and we stashed them in the Nisan Patrol car, heading for Port-au-Prince. We drove through streets lined by collapsed houses and mountains of rubble, teeming with busy or displaced Haitians, past huge tent cities. Lighting was spotty, a faint ghostly blue, and the destruction massive but strangely patchy. The gaily-painted busses, vans and 'tap taps', all stuffed to the gills with passengers and luggage, moved at a snails place. Traffic jams were everywhere. Entering Port-au-Prince, we passed by a huge flattened pile of rubble, the nursing school, with 200 young souls, crushed to eternity by a devastating cave in. The once gleaming white Presidential Palace--of Papa Doc fame during my years there-- slid past on my right, now lop-sided, collapsed, too dangerous for President Preval to work in.

Later we drove past the Episcopal Cathedral, also in ruins. Tent cities were scattered everywhere as we threaded our way along the crowded streets. The driver, Matthew, and I jabbered away in Creole and French, while the other new doctor, Zurob, from Russia, Georgia actually, sat pouring over his tropical medical manual as we jounced over an incredibly pock marked road, strewn with rubble and refuse. I had no clue Zurob was to be the overall boss of Primary Health Care, over me but way up the ladder. The first thing he did in the van was teach me to pronounce his name: " 'Zoo', like the animals, 'Rob' like the thieves."

Because of the traffic and rubble road blocks, we wound our way around Port-au-Prince, passing many tent camps, pitched everywhere. Not all tent camps were the same, however. The ones in front of the Palace and the Cathedral were more orderly and manicured. As I came to understand the situation, these locations represented prime property, coveted by the sponsoring NGOs and the Haitian government. Everyone wanted these prized spots in order to have the limelight on the national and international stage, for media recognition and fund-raising leverage. These were laudable reasons for the most part, and reflected a very particular pecking-order among NGOs. Those with more clout got the prime real estate, wanting to put their best footprint forward. First estimates were putting Port-au-Prince damage at 70 to 80 percent.

When I arrived at the Plaza Hotel, I found the main rooms filled with cots and mattresses, eager excited exhausted young physicians from all over the world coming back from a day's work in one of the remaining hospitals (Port-au-Prince's General Hospital HUEH), and outlying clinics.

I felt anxious and out of place, no one at first greeting or orienting me—until I wandered into the dinning room, sampled a sumptuous buffet and sat down at random with a group of doctors. Everyone was talking about life and death, and the resilience of the Haitians, and the malingerers and those not really acutely sick but wanting a doctor for old ailments. Triage, compassion and breathtaking work were in the air, as they began to clue me in about how awesome the effort was, how great the support from the leadership of the International Medical Corps was, several showing me where to find a spot to sleep, telling me about cornflakes for breakfast, about lights out at 10, on at 6, bus leaving at 7 for the hospital, AND, where to find the WiFi, which was good.

I slept my first night on a real bed, but pulled out the mosquito net contraption I brought with me, pulling it over me like a cocoon. Surprisingly, I slept well, waking fairly early. It was still dark. I grabbed my flashlight, groped my way into the bathroom, needing to crank it. It had no batteries, required charging by hand, and made a lot of noise. As I was brushing my teeth, I heard someone moaning. In walked this sun-tanned guy with a blond beard, gingerly holding his left wrist bent at a gruesome angle--fractured and badly dislocated. I could hardly look at it.

"Good god, man, what happened?" I said.

"Fell down some stairs in the dark just now. Know anyone who can help me with this?" I looked at it and shook my head. Being a psychiatrist, I was a long way from setting bones.

"You with IMC?" I said.

"Nope. I'm on my way out to Partners-in-Health Hospital. No one from there is here now. The desk said IMC could help."

I was about to say no, but then had a better idea. "I know just the guy. Zurab."

"Zu…who?"

"He's our chief medical guy. Wait right here." I edged my way between cots and mattresses back to where Zurab was sleeping, woke him out of sound slumber, and told him about a Partners-in-Health guy in the John. It took him a while to get the picture. Anyway, I had seen my first patient, done triage, and made a referral, all by 6 am in the morning, and he wasn't even Haitian.

I eyed the hotel pool eagerly. Security was good in the compound, not good on the streets, the staff being like mother hens brooding over us volunteer chicks around safety issues. With the great hotel buffet for all meals I thought my fear of losing 40 pounds like before was history. I had landed in a gourmet palace.

Saturday, March 6: Port-au-Prince: Hospital Shock

This was not to be for long. The next morning I met Nick, the British psychiatrist taking me under his wing, with driver and a Haitian electrical technician transporting a generator, along with Mark, a bright eager translator who stuttered. With salt and pepper hair well cropped and coiffed, Nick greeted me with a warm smile, putting me at ease right away. We set out on morning rounds which lasted all day, with no food--because of the unexpected, at least for me. We went to

WHO's PROMES facility, a huge warehouse complex, to get psychiatric meds for an outlying chronic psychiatric hospital, Defile de Beudet, in Crois-des-Missions where we would be visiting next. Procuring the psych meds ran into red tape, cut eventually by helpful French staff, but still taking three hours. With our supply in hand, we were off to the hospital somewhere beyond the airport. I finally realized it was close to my initial field site 50 years ago where I caught malaria. When I told Nick about riding to the Port-au-Prince hospital on top of pigs, he looked at me, "Come on, Kent, you must have been hallucinating." Everyone laughed.

As they flashed past, the caved in buildings reminded the electrician of the stench of bodies, saying we were lucky it had abated through decay and rats, and then desiccation (that drying out process). He added that the approaching rainy season might stir things up again. He was stoic and fairly closed mouthed about all this, until he sighed deeply as we passed a collapsed building. A dear friend was still buried there beneath the rubble. His family tried to find him, picking through the smaller pieces of rubble, but hadn't found him yet. I noticed a single tear roll down his cheek. We were all quiet for a moment, until he resumed his patter.

The traffic was incredibly bad, but we finally reached an alley running between caved-in buildings leading on through stands of banana trees. We splashed over cess-filled ditches that only the dreaded rainy season could wash clean, until mudslides buried them again. All the while, Nick led us onward, briefing me about our teaching mission with the young Haitian medical professionals in our IMC mobile clinics. We were also providing consultation, teaching and support to the remaining Port-au-Prince mental hospital, as well as the IMC's outlying medical clinics. He was wonderfully British, warm and supportive, inspiring a bit of hope and confidence in me even as my head swam. I was trying to imagine myself working on the front lines, in the trenches, under such difficult and compassionate circumstances, wondering how he kept his cool.

I finally mentioned my misgivings, which brought a quick smile. "I've seen your CV. I think you'll squeak by." He had a knack for leavening his sober realism with quiet enthusiasm and a wry sense of humor.

Finally we reached a metal gate with a shotgun-toting guard, who let us through after brief questioning, revealing a huge open space pockmarked by caved in one story buildings and tumbled walls ringing the grounds.

The inmates couldn't sleep safely indoors any longer, and now lived in tents. Touring the grounds with the administrator, Mr. Aubin, we inspected their collapsed primitive kitchen, and marveled at their tent city.

We discovered that when Mars and Kline hospital in Port-au-Prince released most of their patients after the earthquake, many came flooding out here, where some had been before. Any port in a storm. A patient with an American flag bandana was screaming and gesticulating at us, while two women stood frozen in a bizarre embrace peering at us.

Men were marching about just behind us, half naked women lay sprawled on tattered cots, or slept on the ground, while goats fed and frolicked about. Nick toured with the second-in-charge, doing a 'needs assessment' as I trailed behind saying hello to all the curious patients. I was surprised to meet one who spoke English, telling me he had just won a Port-au-Prince art award for his wonderful drawings. He wanted to give me a portrait for helping them.

We delivered the huge supply of meds, my friend the electrician installed the generator for their water supply, and the heavens opened up on us, albeit briefly. One of the frozen women untangled herself to say in broken French that they feared their gleaming white canvas tents from Russia were old stock and might soon leak. When they saw us taking pictures of the staff and director, they asked, then insisted, we take their picture too, just for me to have privately.

When I got back to the hotel I was told to pack up for transfer to the Residence in Petionville, where the staff of IMC lived, to get to know them better. It would be easier to orient me for mobile medical clinic work and the seminar teaching I would be doing, and closer to the supply of teaching materials and advice.

"What's this all about?" I asked.

"We're getting you ready to assume leadership in the outlying clinics in Petite Goave," Lynne said.

I looked on the map and realized Petit Goave was just beyond Leogane where my original field site had been.

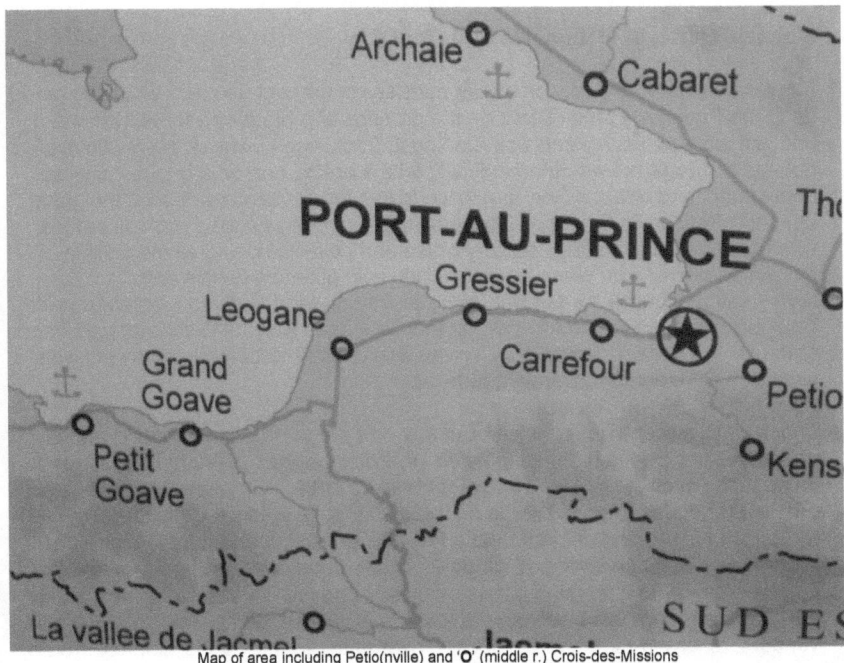

Map of area including Petio(nville) and 'O' (middle r.) Crois-des-Missions

My beloved Leogane had been the epicenter of the January 12th earthquake and major aftershocks.

"The last tremor," the translator told me, in between stutters, "was over a week ago, beginning with a huge snapping groan, followed by a slight convulsive shake." Nick told me that when one of the aftershocks hit, everyone moved outside the Residence to sleep, fearing building collapse. Most stayed out there for some time. Some of the surgeons and nurses got freaked out, insisting on leaving Haiti immediately.

Because beds were in short supply and I was transient, I was assigned to a tent outside because I was already slated to depart next Thursday for Petit Goave. Today Nick and I headed for the Port-au-Prince Hospital Psyche Clinic so I could warm up my skills, medical and psychiatric, as well as my Creole. I felt inspired, daunted, and challenged, and glad I was here. This is an amazing country, being helped by impressive groups and dedicated younger people. I wondered how I would hold up physically with all the pressure and the heat. And, after my nightmares on the plane, how would I do mentally? I knew I would try hard as I put my ancient oar into these troubled waters. May the good Lord and my ancient friends, the Voodoo gods, look kindly on my efforts.

Monday, March 8: Petionville Club Tent City

I awakened early. As I lay on my air mattress in the tent, I knew I would be going up to the Petionville Mobil Clinic on the grounds of a sprawling tent city there. I had just learned Sean Penn was running it. Back when I was doing my fieldwork outside of Leogane I would periodically take a break, hop on a brightly colored camionette, and head to Port-au-Prince to join the Richardot family, often going up to the Petionville Country Club where we swam or attended cocktail parties, swarming with Haitian Elite, military brass and politicians, and all the visiting NGO dignitaries. The place would be filled with glittering jewels and conversational gems, awash with five-star Barbancourt Rum and festooned with Bougainvilleas. As I've mentioned, at that time Jean Bleyfus Richardot was head of the United Nations Economic Commission to Haiti, accompanied by his wife, Natalie, and daughters, Carole and Nancy.

Those were sweet times of relief. But this was not to be in the morning that lay ahead. After a quick breakfast, I joined Dr. Peter Hughes, a bright intense ruddy-faced Irish psychiatrist. We worked our way up to the Club, passing a hillside of destroyed houses. He filled me in on the IMC mobile medical clinic there, staffed by Haitian nurses and doctors, and a few IMC physician and nurse volunteers. We turned into those familiar Club gates, some of the metal letters now dangling by one screw, into a teeming jungle of military and NGO vehicles, camo-garbed gun-toting marines, blue-scrubbed doctors and nurses, and gaudy-shirted Haitian vendors, all plying their trades. I looked over at the tennis courts, one covered by a behemoth tent covering a dozen smaller tents, forming a small orphanage compound, supplies stacked around the periphery.

The other two front courts were pock-marked by tire tracks leading to the back courts used for army vehicle and a supply depot. With apprehension and curiosity, I walked out onto the Club terrace, noticing cracked columns, and huge jacks holding up cement cross beams, camo partitions and windbreaks obscuring my view. The NGOs (non-governmental organizations) including us, were gathering on the right, and the army was on the left.

These guys were all huge and buff in their camo fatigues, guns leaned casually against the wall, some lounging in front of a huge TV watching March Madness (that American basketball tournament). Zombie-like I walked forward to the edge of the terrace, looking out over the empty pool at the withering Bougainvillea, one branch bearing faded red flowers. I could make out three NGO tents on the crest of the hill beyond, their names emblazoned on their sides, and a few army vehicles. Stepping down debris-littered steps to the pool below, I could see Port-au-Prince and the mountains beyond shimmering through the mid-day heat. The Club's former grandeur, with its panoramic view, was still fadingly evident.

Peter, Kettie our superb Haitian psychosocial nurse, her new assistant, Margery and I had a pre-arranged rendezvous with a Canal 24 television reporter and her video cameraman. We were their entrée into the camp below, and they were possible resources for stirring donations to the Haiti relief effort. And long-term relief is essential. Bright and attractive, our reporter joined us, and we walked together over the brow of hell and suddenly were confronted by a breath-taking sight, the multi-colored mosaic of a tent city now housing 50,000 Haitians, spread out like a cubist painting, over the contours of the hilly golf course, ravines slashing the landscape like tentacles.

Our group had its clinic nestled in the middle of this tent city, with other NGO care and health groups scattered around, most placed strategically near the periphery. We threaded our way down into this warren along dirt paths, zigzagging our way through tent alleys, the Haitian tent 'homes' neatly divided by blankets inside to provide some privacy, with life spilling out into these walkways, kids playing, and mothers nursing and cooking, and men and women selling their wears every few feet.

There was civility and curiosity at every step, eyes following our descent. I often stopped to say a brief hello in Creole, always receiving a smile and polite hello back. I took a picture of two girls combing the hair of Barbie-like dolls, and a bunch of boys with kites made of refuse, one flying quite high.

Suddenly a long string trailing behind a little boy caught my eye, with a piece of red refuse at its end. What was he doing? Then I caught on. He had a red puppy on a leash. I yelled when Peter almost stepped on it, saying, "Watch out for the puppy!" The boy and his friends cracked up. The child psychiatrist in me was ever on the look out, trying to take the pulse of these kids. I was struck by their resilience and strength, even in the midst of tragedy. If their parents, the government, and the NGOs could provide shelter, water, food, community and security, they could adapt and thrive. The mantra of IMC was hitting its mark, at least until the monsoons or a hurricane hit harder. Then there would be hell to pay. I banished the thought, smiling about that red puppy.

We spent several hours at the clinic seeing psych patients and working with one of the young Haitian doctors around improving his front line psychiatric skills. He in turn taught us about what he was seeing medically. We all grew, and were in awe, as we sweltered together in the hot tent, seeing our own psych patients, and watching the other doc grinding out about 70 to 100 patients over several hours. The Canal 24 videographer took shots of the patients and staff, capturing the long lines outside, and the extensive tent city teeming with Haitians.

They asked Peter to come out at one point and interviewed him for about five minutes. When he came back in, to my surprise they asked me to step out for an interview. She explained their station was English speaking and that interviewing an American would have some cache. I explained I was fairly new to the IMC. She countered that I would bring a fresh American eye to the whole scene.

After the interview, lasting 4 or 5 minutes, she said, "You're quite articulate in front of a camera for an IMC new comer. Have you ever done this before?"

"Yes." I said, "I was live with Tom Brokaw for the Challenger Disaster and on the Diane Rehm show the morning off the Oklahoma Bombing. Oh, and the Charlie Rose show."

"No wonder," she said, "I've heard of Tom Brokaw. This will be shown in Paris and all over France sometime soon."

When I walked back into the IMC tent, I down played it, because I thought only Peter should have been interviewed. Later Peter let me know we were both shown on Canal 24. He seemed happy to share the limelight.

Finally, we walked back up the hill past two huge American army guys, in camo outfits, machine guns kept discretely low or out of sight. I asked to take their picture, and they smiled knowingly, lowering their guns and assuming an at-ease pose. Seems they've been asked to be photographed before. Then I asked Peter if he would take mine, and then returned the favor.

For me, the transformation of the Petionville Country Club was mind boggling, stirring confusing emotions of awe and disbelief. I felt an upwelling of respect, appreciation and hope mingled with anger at the earthquake. Sadness for Haiti's already impoverished condition already formed a backdrop to this chaos. Stunned by this scene, I worried for her difficult smoldering future.

An email from my friend, Edward Hughes, the night before had opened my eyes. Because, if the Haitian government, despite all the help, took too long to pull itself out of it's understandable disarray, there would be trouble. To set this in context, let me quote something that Edward said, "I was chatting with my daughter Hannah today about Haiti. She is co-authoring an article with one of her professors at the London School of Economics on confrontational civil disobedience in Africa, or something like that. In the course of her research she has extensive contacts with OXFAM. She informs me that the OXFAM people are worried about various scenarios in Haiti, any of which might result in the complete collapse of governmental authority and a surge of wide-spread, violent civil unrest. Apparently the consensus is that the situation may well deteriorate quite rapidly over the next several months as the country descends into further chaos and a violent upheaval might take hold.' I convey all this because Hannah is not given to hyperbole (unlike her father). When she says, "Watch out", you are advised to duck. You may have seen the beginning with the kidnapping incidents, but it is going to get a lot worse. Do you have a weapon?"

Because Edward wrote so well, and lived in Haiti as a young man, let me quote further: "Living in Haiti in the early '70s I learned that notwithstanding the wretched poverty of its people, the island is beautiful: brightly colored bougainvillea set against a deep blue sky over an aquamarine ocean. I would wake up in the morning looking out over Port-au-Prince from the wide terrace of our Petionville house and exclaim, "Another goddamn beautiful day!" I often wandered down to the port or around the old center, littered with remains of the 1915-1934 Marine occupation: the rusty lamp posts erected on the side of the road, now oddly standing not quite in the center of the widened street; the rotting porches and facades of gingerbread structures, side by side with tin covered lean-tos. I was both compelled and repelled to hear, see and smell the chaos that is Haiti. The smell, in particular the mixture of diesel, rotting fish and open sewage would wrench the strongest stomach into quivering ectoplasm. The stark contrast between what I saw and what my other senses felt created a dissonance I could not reconcile, even with measures of cannabis and Barbancourt. It is a place of strange extremes. Earthquakes don't erase the sky or kill the flowers, but I imagine the squalor overwhelms what meager solace the sky and sea provide."

At breakfast the next morning I was very sleepy, kept up by very disoriented roosters crowing randomly all night. They each had signature calls and seemed to incite each other, sometimes setting off cascades of crowing ricocheting all around me. I got to know their individual cries, and thought of purchasing all of them for dinner the next day. I also suspected that Edward's daughter's dire predictions were weighing on my mind. I knew for whom the cock crowed. Living in a tent, though quaint and charming, was a bit cramped, making my morning ablutions and stretches awkward, punctuated by escape of morning noises, with sudden recognition they were now a matter of public display. Tent walls created an illusion of privacy. Stiffness was my constant companion at my age, and at first I curtailed my regimen, a concession to shyness the first few tent days, until I said to hell with it and began to do my stretches vigorously (and noisily) again. I felt better for it all day long and no one complained. They were either asleep or enjoying the noisy entertainment.

It was amazing how chaos could reign in such a small tent space. As I fell asleep, I slithered out of sweat-dampened clothes, dropping them randomly in the dark, exhausted. I was degenerating slowly into a disorganized minimalist, having brought in an embarrassing amount of stuff. Why were Conrad, and his 'Heart of Darkness' floating repeatedly through my mind? When my daughter, Julia, heard I was living in a tent, she said, "Daddy, you're too old to do that!" But my son Christopher, who has camped a lot with me, said, "Ah, Dad, you'll do fine. Your Boy Scout training will really come in handy." In the dark of night, though, I felt a little doubtful at times. Then the stark rending images of the vast tent cities and the homeless filled my mind and I realized how fortunate I was, with all my high tech equipment. Sleeping in a tent put me a little closer to my job--closer to the people we were serving.

Yesterday, I went with Nick as his sous-chef to see the IMC facility at the Haitian University State Hospital (HUEH), still mostly under Haitian control.

The IMC triage, outpatient, ICU, and medical units are in tents, staked around the collapsed or unsafe buildings. We visited the ICU first, to see a 30-something woman looking good despite periodic rectal bleeding and a paralyzed left arm and legs. This came on unexpectedly many days after the quake. For such dire straights she seemed surprising comfortable, exhibiting classic *la belle indifference*. We were asked to examine her because the distribution of paralysis and the history didn't make sense, nor did the bleeding pattern. We felt she might be having an hysterical conversion reaction, expressing her emotional conflicts through bodily reactions.

But then I thought about the bleeding pattern, and said to Nick, "You know, she might be a Munchhausen Syndrome, faking her symptoms for emotional and social benefit. She's a nurse and would know what to do medically."

I asked if anyone had actually seen her bowel blood, or was it just claimed by history. It turned out they had seen it once, at the beginning. I had been the regional Munchhausen expert back in Washington, DC, and had seen patients bring in hidden sacs of blood to put in their bedpans, to get or keep themselves in the hospital, so they could be fed and taken care of and get away from horrible outside situations. If she was doing this, we just didn't know how. When making this diagnosis, I always fear I might be making a mistake and missing something physical and serious. We told the nurse and doctor our thoughts and said the patient should be transferred to the step-down medical unit because she was taking up valuable urgently needed space and was getting too much reinforcing attention. She hadn't had rectal bleeding for several days, but I warned it might reappear when she heard about being transferred.

Shortly after we left she had a significant bowel movement with bright red blood, even though her proctoscopy right afterward was negative for any source. So they didn't transfer her, and gave her an indwelling venous catheter, to give her blood if urgently needed. I suggested the next time it happened they should compare her own blood type to the blood in the bowel movement, fairly certain it would not be hers, though I could be wrong. And with an indwelling venous catheter, she could easily get her own blood to put in her bedpan. But people usually follow their usual pattern. Also, she was still having her paralysis. After our follow-up visit, as we were leaving and she thought we weren't watching, she was talking to her sister and we noticed her making normal left hand gestures during the conversation using the 'paralyzed' arm and hand. We felt our diagnoses were virtually confirmed and were encouraged she would recover.

The next patient had lost his leg below the knee when his collapsing house crushed his lower leg. He was recovering from shock, still speechless and stunned many weeks after. A relative let us know he had been taciturn and depressed for some time before the quake, so we felt anti-depressant, anti-anxiety medication would help resolve his stuck state of mind.

In the next tent, down at the end, we were greeted by an animated young woman in a Bob Marley T-shirt, chattering away in fragmented English and singing American pop songs at us.

"You need to let me go home; there's nothing wrong with me; I feel great. Everything is fine. Ask my husband." Her 4-year-old boy peered out from behind her, tugging at her T-shirt. Her husband, slumped on the bed, shook his head.

"How was your night?" asked Nick.

"She began to calm down and actually slept a little," said her husband. "Thank god. I'm dead. Oh, and she ate dinner."

Nick and I looked at each other and nodded. We both knew she was beginning to cool down a little with an anti-mania drug, which he increased slightly. Physical things like return of sleep and appetite often reappear before overheated mental functioning slows back down toward normal. Her judgment would soon return with sobering impact. Her manic episode was precipitated by her house collapsing in front of her, injuring her mother badly—something she would have to begin to deal with emotionally once she calmed down.

I had noticed a slight elderly woman curled up on the floor directly opposite our manic patient, who wouldn't look at me. She was excruciatingly shy and suspicious. Only once did I see the slightest flicker of a smile—when the 4-year-old peeked around and tugged on the Bob Marley T-shirt. I found out she was our next patient. This quietly sad woman, with her private paranoid delusions, was also homeless. She had been found naked walking the streets of Port-au-Prince. *Oh, my god*, I thought, *Could this be the very woman Lynne and Peter had quizzed me about by Skype in the Dordogne?* As we walked away, I asked Nick if it were so. "Kent, I don't know about you sometimes," he said with a wink. "Let's stick to our patients. The timing's way off anyway. That call to you was a long time ago. Anyway, she's paranoid and needs an anti-psychotic medication, though maybe a traumatic experience set it off. Once she's better, we need to reconnect her with her family and community to secure her recover." I was beginning to catch on. Disaster psychiatry was much like military trauma psychiatry, trying to treat the acute symptoms fast and then get them back to their unit and their buddies, who would bond and talk them back health.

We moved on to an isolation tent, overheated and stifling because of protective partitioning and the mid-day sun. Inside was an emaciated elderly woman with suspected TB, sitting up repeatedly and spitting into an ivory bedpan. I was reminded of my visit to Bonheur in Brooklyn. She came to our attention because she was hearing god speaking to her in the center of her head, beseeching her to be more faithful. With rest, hydration, food and medication, she had slept better, though fitfully, so we could talk with her more coherently. The day before her daughter had said her story was jumbled and garbled, but now she was better, talking more understandably. The medication Nick had used was a mild sedative, and her emerging coherence proved her not psychotic, suffering more from a physical or organic confusional state, now resolving.

Finally, we visited a woman who also saw her house collapse. Initially Nick thought she was psychotic but today we got a better history pointing to psychomotor epilepsy, plus lab results revealing a severe anemia. Even when someone is not actively having a seizure, underlying abnormal electrical activity in the brain can create abnormal mental states. And the severe anemia with resulting abnormal heart rate and brain oxygenation, along with other related metabolic effects, can also cause mental symptoms. Anti-seizure and anemia medication, along with careful follow-up by our team were indicated rather than anti-psychotic medication. The two Haitian nurses rounding with us were bright, gracious and serious, giving the psychiatric meds and supportive follow-up therapy. I found my Creole wanting in these situations, and needed their help, aided by our eager translator. For all his eccentricities, he was a great guy, very bright and helpful. I also soon realized, as my Creole improved, that some of what I thought was stuttering was almost a literal translation of Haitian Creole speech. He stuttered a little, but much of his repetitive way of speaking was due to literal translation of what some Haitian say in Creole. He talked the way he thought. And I slowly learned something.

By this time I was famished and thirsty, ready for a mid-day break and at least one of the Power Bars I had stashed away. Actually I had several and planned to share. But that was not to be. We were late to get over to what I mistakenly heard as the Calvin Klein Hospital (actually, Mars and Kline), where Nick, without missing a beat, delivered a great lecture replete with interactive exercises for the assembled roomful of Haitian psychiatrists and psych nurses.

We entered the Mars and Kline compound, situated in the center of Port-au-Prince and surrounded by a high wall to keep people out and protect the privacy of the patients, I was amazed to discover the grounds full of tents bursting with Haitian families, all engaged in normal Haitian life, peeling vegetables, selling wears, watching soccer on jury-rigged television sets, tending babies, all living on top of each other right up to the hospital door.

The buzz and bustle of life was everywhere for these lucky ones who found such a choice spot after the earthquake. Mars and Kline had thrown open her doors to the great unwashed, releasing their healthier clients, retaining a few hard core patients, and hosting the displaced citizens of Port-au-Prince.

During Nick's lecture, the buzz of communal life created a background hum, drifting in through the open barred windows. In the middle of his lecture a horrific metallic clanging began, and persisted, punctuated by loud screams. I was quite distracted by the Bedlam, but no one else seemed to notice. This went on periodically throughout his lecture and exercises. Nick didn't raise an eyebrow. Afterward, we went on a hospital tour and I found out the noise came from the few remaining inmates, many in isolation cells. They were tuned in enough to know when to beat their metallic tom toms to torture a captive shrink audience. Actually, they wanted attention and better care.

These docs and nurses at the conference participated with enthusiasm. And I was impressed with the turnout. They jammed the room, putting chairs at a premium. Times were difficult and money scarce, though. So how did they find time and interest? Then I discovered a partial answer, over and above the great lecture. At the end, white styrofoam boxes began cascading into the room, delivered by our driver. Everyone was treated to a great Haitian box lunch *grace a IMC*. What's in a classic Haitian lunch? Baked chicken with that delicious piquant red palm oil sauce, crispy fried plantain, and savory rice with red beans. I was famished and waiting for mine--but there were barely enough and Nick alerted me, "IMC family hold back." Actually, it turned out he didn't think we should eat in front of the group, and on the IMC nickel. I was beginning to feel lighter already, not just from steady sweating. Nick was slim and trim, and I had a number of excess pounds anyway. Maybe I would be losing some weight after all. And I felt less guilty about all the starving Haitians surrounding us outside.

That night, after my troubled sleep the night before, and a string of early morning awakenings, I figured I didn't need to set my alarm clock. "Kent, you in there?" Nick's voice penetrated the tent like a knife, awakening me. For once I had slept in. It was 7 am, and we had made a deal. Here's the background. With all the security restrictions getting tighter, I said to Nick at breakfast the day before, "I'm already going a little stir-crazy, and I haven't been here that long. I never imagined it would be so locked down like this."

"How was it when you were here before?" said Nick.

"You mean fifty years ago? I roamed all over Haiti at will. I'd take off and go anywhere by camionette, tap tap, or foot—even with the background specter of Papa Doc's ruthless Tonton Macoutes (his private police) shooting people. Somehow that just didn't bother me back then. I figured I knew the Haitians, could speak Creole, was a young student, and would always be okay. Of course, I also had a letter of 'Sauf Conduite' (Safe Conduct) stashed in my pocket, signed by the Haitian government. Now our own security is watching me and I can't even go outside the building. Seems like I can't even take a leak without someone watching. It's getting to me."

Just then several staff members trooped in, hot and sweaty, led by Charles, our tall, muscular logistics guy.

"Where you guys been?" we said in unison.

"A long run up into the Petionville hills through the subdivision behind us here. Great views up there."

"What about security?" I said.

"Too early for the bad guys to be up, it's broad daylight, and we always go in a group." Hearing this, I just shook my head, looked at Nick, and we hatched a plan to run the next morning.

When I heard Nick's voice through the tent wall, I said, "I'll be right out". But to myself, I thought with relief, *to hell with my stretches for once!* We headed up a dusty road between pastel houses, some caved, but many surprisingly intact (apparently they used real cement in this high rent district), winding to the crest of the subdivision hill. From there the view was indeed spectacular, out over Port-au-Prince, the Airport, and Croix-des-Missions, a plain spreading out to the right back toward a lake (was it Etang Sumatre?) its waters shimmering in the early morning heat.

"This is amazing," I said. "Over there is Saut D'Eau where I went with Coyotte, my drummer friend, to see all the Voodoo possessions during the waterfall festival. Talk about freedom to explore!"

"Must have been extraordinary," said Nick. He wasn't pushing the party line this morning. He loved getting out as much as I did.

Then I looked down to my far left, and got quiet. Nick saw the wind go out of my sails. "What's up?"

"Down there somewhere is Lilavois. That's where the orphanage and school are for the little 3-year-old I sponsor through Haitian Outreach, a Massachusetts child sponsorship program. Her name is Makenta Paul. I'm dying to find out if she survived. I hear the area had a lot of damage, though spotty. Things are so tight at IMC I wonder if I will ever get there? The two things I promised myself I'd do while I was here were a visit to my field site to see my friends, and a trip to Lilavois to see little Makenta."

"I hope you can. It may be dicey. So play your cards right. Just give IMC time to get to know you, and see how the security situation goes."

At breakfast, after our run (lots of fast walking, actually), the head of the IMC Outreach Team sat down with us. Alice was a young French woman with fine aquiline features who had first worked in Afghanistan, then in Iraq, and now had been among the first to arrive in Haiti. She asked each of us why we were here, trying to see where we fell on her 'motivation triangle' of 'careerist', 'adventurer', and 'tree hugger'. I was moved by each story, and found the logistics guy wise when he said 'tree hugger' wasn't a fair designation for liberal idealists who should be respected, provided they have some realism too. He then mentioned his own triangle terms: 'Mercenaries', 'missionaries' and 'misfits'.

Nick and I made rounds at the hospital, discharging the Bob Marley manic girl. I turned around, and to my surprise, the quiet little lady smiled at me, just a flicker. She was getting better too. Next we found out the suspected Munchhausen had been transferred to a quieter step-down tent. One of the Haitian nurses had overheard her whispering to her sister about whether her bleeding could get her transferred to the Comfort, and sent to United States. Indignant, the nurse told the resident, convincing him she was manipulating the system like the psychiatrists had said. We felt it best not to reward her with a visit. She would get tired of the ruse and get herself discharged.

We went on into a part of the old hospital that was still standing and safe to do a follow-up on a 56-year-old woman. She had seemed so severely depressed, almost catatonic, that we had 'hit her fairly hard' with an initial starting dose of Amitriptylene, a tricyclic antidepressant particularly good for immobilized middle-aged depression of this sort. But there was something about her that made us both worry we (and her Haitian doctor) were missing something physical, maybe a brain tumor or a blood clot. She had already been given heavy doses of anti-psychotics, making us also wonder about neuroleptic malignant syndrome (a feared but rare side effect). But she wasn't very rigid and not at all feverish, cardinal symptoms of this syndrome.

We only had two hours before the next Mars and Kline lecture, so we swung into action, setting off on a 'mission impossible', trying to find a working MRI or CAT scan machine to do brain imaging some where in Port-au-Prince. The hospital had none. Networking, we discovered the address of one and shot over. The doctor was articulate and willing, except for one thing. His scanner had been damaged by the earthquake and wouldn't be on line for a week (or two or three). He was drumming up future business. Finally he told us of another machine, owned by a competitor, and off we went, our driver sure we were on another wild goose chase. This machine, after a long collegial discussion and presentation of our case, proved to be intact. We settled on a price, Nick and I pooled our resources, and forked over the money. But how to get her there? I had chatted up an ambulance driver at the hospital the previous day when I had a minute to kill. I called the Norwegian Red Cross guy and talked him into transporting her to and fro. After the Mars and Kline lecture, we found out that for once things had gone smoothly. The brain scan showed no evident pathology. We were relieved but mystified, and back to zero. Unfortunately, later that week she died, leaving everyone with big sad question marks. We had tried our best.

Peter, our ruddy Irish psychiatrist, delivered another fine lecture at Mars and Kline, punctuated by the same staccato inmate clanking. As we left the lecture, Peter walked up to a young woman with a cameraman in tow. She was a reporter from the New York Times. Though he professed distrust of reporters, he clearly enjoyed this one, and knew how to work her (and she knew how to work him). They were allowed into the inner courtyard to interview and photograph at will. I bootlegged some pictures too.

The acting director, Dr. Franklin Normil, wanted any help he could get to raise funds and supplies (he knew how to work everybody). The young boy was still lying nude on the pavement, the same nearby residents clanging away for attention (and a photo op). We were at the near end of the courtyard when a horrendous banging began at the far end, from someone I met on the previous visit—a big black gentleman by the name of Napoleon, scary at first with his scared face, but with a ready grin displaying a missing front tooth offset by a shiny gold one. He looked like a cross between Sinbad the sailor and Scarface. Last time he told me a benign story of mistaken identity in broken French. I suspected otherwise.

I sauntered back, saying, "What's up?"

He replied, "Hey, look, I'm actually an American citizen and deserve equal air time. If you get that beautiful reporter to come back here, I'll give her what she wants."

"I'll ask her." With that he stuck his massive fist through the bars and gave me an Obama high-five knuckle chuck.

I invited her, "Come meet a fellow American," and she obliged.

"Why are you in here?" she asked, "And who chewed up your face?"

"It's my fifth time. I drink a little, snort a little, and get high as a kite. I fight a little, and the police don't like me. So they stick me in here."

"Look," I said, "that might get you in jail, but not here."

"Okay, you're right; maybe it's the other way around. I start getting really high, my mind races, I get these fantastic ideas. Think I'm superman. And I medicate myself with a little booze and smack. Each time I forget the bad consequences. I buy cocaine, I sell it, maybe a few guns on the side. I need cash to feed my habit, buy my booze and heroine. The police finally figured out I need to be in here for some doctor drugs to cool me off. Because of the earthquake I'm stranded in here. They don't want to let me loose. Can you get me out if I promise to be good?"

The reporter's eyes got big, then narrowed. "You telling the truth, or just jive talking me for the lime light, buddy?"

"I would never lie to a pretty girl like you." The group was leaving so we had to break off and hurry back. The reporter stayed to talk with staff, but we were late for the clinic.

Just outside the hospital wall, there was a huge refuse pile. Lying on top was an emaciated old man with a scraggly white beard and gnarly toenails, completely nude except for patches of caked mud. He lay there motionless, his eyes glazed, only his eyelids flickering occasionally. *He looks catatonic, and he's on the wrong side of the wall,* I thought, *or maybe he should be in a medical hospital for some dread disease and dehydration.*

"Don't touch him," said Peter. "We can't get involved like that out here. It's dangerous to be a good Samaritan except in our protected clinics. He may be bait. If you straggle, get separated from our group, and pay too much attention, you may get accosted one way or another."

"Really?"

"Yeh, somebody will pop up and say taking pictures steals his soul, or worse, you're violating his rights, and demand money."

Interestingly, when her article appeared, the New York Times reporter noticed this nude man too, mentioning him along with Peter, with a nice mention of Lynne and Nick, giving good objective coverage to the International Medical Corps. Back at the Residence office, Nick gave a summary of what we were doing and then Lynne began to do some teaching nearby.

At breakfast the next day, I was refreshed by the evident idealism at the table. But that unfettered idealism was dramatically chastened when someone came in breathless announcing that two women from *Medicins Sans Frontiers* had been abducted the night before and everyone in the NGO community was waiting for the ransom request to come in. I began to take Lynne's security warnings much more seriously. A previous abduction ransom request a month ago had started at a million and was settled for $40,000.

The table discussion shifted to group security management for the new IMC volunteer group just arriving. The night before they had already received a sobering orientation about working in a destroyed city with dysfunctional hospitals, alerting them that despite all their expertise and hard work they wouldn't have the resources they were used to at their disposal. There wouldn't be the usual support services and medical and surgical specialty groups. They should prepare for frequent frustrations and deaths despite their best efforts. But after these caveats, the orientation had finished on a more encouraging note. "At the end of each day, despite all this, you can come back here having had a very special, rewarding experience. You will have done immense good with scarce resources serving a grateful Haitian people."

But now, here we were as a staff group, having to deal with the abductions and the few bad actors out there. The discussion turned to how the new Volunteers should be alerted and dealt with. I looked at Nick across the table. Our eyes met, and I could tell we were thinking the same thing. We were glad we had gotten our morning run in the day before. We sensed security restrictions would be tightened even further.

On the way home from evening meeting where the volunteers had their security brief, I praised our young Iraqi IT specialist volunteer for creating our excellent internet and Wi-Fi system at the Residence. Then I asked him about other places, like Petit Goave where I was going. He assured me he had made certain Petit Goave would be the same as Port-au-Prince even though a more remote outpost.

"How about the drive out there, what's the road like?" I asked.

"Take a look at Google Earth, you can actually see the road in high definition yourself. Part of it was just redone January 23rd, because of landslides. But when you look at Google Earth, prepare yourself for a shock. The rest of the video is a month old and you can still see bodies stacked along the sides of the roads."

My big news that day was finding out from the Director, Lynne Jones, what I would be doing for the rest of the month out there in Petit Goave just beyond the quake epicenter. Part of it would include coming back and forth to Port-au-Prince because of Mars and Kline teaching duties every other week, going right past my field site, my jumping off place at Brache. I thought to myself, *At some point, I'll just make them stop at Masson (Brache) and wait for me so I can visit there and deliver some money I had saved for my peasant friends. Hopefully I can set up a longer visit later to really talk to them and see where I lived.* Though I knew IMC was tight on security, my former anthropologist's sense of invincibility and brashness was suddenly rearing its impatient head. Karen Richman knew somebody there who was running a 'Masson' relief effort for the community. She could email me his Haitian cell phone number. Despite minimal infrastructure, every one seemed to have a cell phone. I promised myself I would be careful and safe about it. I felt I could definitely pull it off. After all, my Creole was coming back, and so was my personal comfort and sense of trust in the Haitian people. At least most of them.

Lynne Jones told me she had designed on paper a mental health clinic, to be attached to a small Haitian general hospital in the heart of Petit Goave, called Notre Dame, now partly run by our IMC group, out on the southern peninsula. This would be a new clinic, not yet functioning, with a Haitian psychosocial nurse and a translator already hired and waiting for me. I would be creating and running the clinic, which would treat outpatient referrals from the hospital's follow-up and out-patient department. Haitian family practice doctors in four nearby outlying family practice clinics would rotate one day a week into my clinic to see our patients and be supervised by me, seeing mental health patients, to train them to incorporate this into their own frontline work. On another day I would rotate out to a 'boat' clinic outpost to do the same thing. Since I would be in Haiti for a month, two out of my four Saturdays in Petit Goave I would be teaching all day Disaster Psychiatry workshops, beginning right away this Saturday. The entire medical staff from all the mobile clinics would come to the Royal Hotel.

Hearing all this really made me feel a little anxious and overwhelmed, especially since the trainees were all Haitian and spoke Creole. Not that I had to do the seminars in Creole. I would have a translator. But nevertheless, it would all be new and complicated. And giving such lectures on topics I had to create always made me anxious the first time. I would be moving out to Petite Goave Thursday to start all this, beginning the Clinic on Monday after doing the first workshop on Saturday. At least I would have Wi-Fi out there, thank goodness, making access to teaching materials and IMC back-up possible. Lynne and Peter had a lot of teaching outlines, and Nick also. My consolation was I would have Sundays off, and Fridays for paperwork and lecture prep. But even so!

My last evening in the Residence I had a pleasant surprise, even if I was fairly clueless. IMC had a bright attractive public relations person by the name of Crystal Wells, who walked in with this young woman. I hardly looked up until Crystal cleared her throat and called my name, telling me she wanted me to meet Sienna Miller, IMC's Global Ambassador for media and fundraising. She would be coming out to visit Petit Goave while I was there. The name meant nothing to me. Crystal ushered her on to meet other staff members. I noticed their eyes all lit up when they met her. They were all cognoscenti, members of some secret society. Not wanting to ask stupid revealing questions, I secretly Googled **Sienna Miller** on the spot. As most of you media mavens already know, she turned out to be a famous English actress, model, and fashion designer, best known for her roles in **Layer Cake, Alfie,** and **Factory Girl**. And she was Jude Law's girlfriend. I had heard my daughter, Julia, and her girlfriends sighing about Jude Law once. But I couldn't place him. I Googled again: "an English actor, film producer and director, one of the most brilliant and unique actors working today. A picture was captioned, "**Sienna Miller** and **Jude Law** stayed lip locked all day while vacationing together in Italy."

Wednesday, March 10: Petit Goave: Unexpected Sadness

With Edward's words echoing in my ears, I found my time was up in Port-au-Prince. So I packed up early two days ago, actually beginning to miss my little tent that had come to feel like home, all cozy and now nicely organized.

Even a rooster came crowing and clucking behind my tent to wish me goodbye. I took pictures of Pierre and Carmen and Alice who fed us *'di ri ak poi rouge'* (rice and red beans), *'di ri ak jon jon'* (rice with French peas and jon jon flavoring), overcooked chicken, and *'banan peze'* (flattened fried rounds of plantain), among other things.

We spoke Creole together, which pleased them. I said my good byes to staff, and followed my driver to the IMC range rover. I had been told that the drive to Petit Goave on the Southern peninsula would take 4 or 5 hours, but he told me 2. Just driving to the Gressier Clinic the previous day had taken 1-1/2 hrs. I was dubious.

We wound our way down the mountain past all the chaos of UN blue t-shirted Haitian conscripts working to clean the debris from the sides of the roads, where everyone was piling the rubble from their shattered homes and businesses. Since everyone was desperate for work, the UN provided lunch and a few dollars and a minimum wage. We hit a few areas where there were bulldozers and dump trucks lumbering about. Moving slowly, they cleaned up the never-ending debris, moved by wheelbarrows to street's edge, or lifted by tyranasauric steam shovels feeding steadily on the carcasses of houses and buildings, repeatedly disgorging their indigestible loads into a line of huge waiting dump trucks. Many of them were ancient Mack trucks from the '50's that the canny Haitian mechanics kept alive. Because our office and Residence were in Petionville, a more well-to-do area above Port-au-Prince, home of the elite and many government officials—and many of the NGOs—there seemed to be more debris removal going on than down below in Port-au-Price. It was also a safer, more secure area, with a larger proportion of better-built, earthquake-resistant houses and buildings. Even so, there was a huge amount of damage and much work to be done.

We worked our way down to Port-au-Prince, and finally crept into Carrefour intersection, amidst teams of brightly colored, Jesus-hailing tap taps and camionettes, and huge Camions, each with their own pictures of Jesus or pop stars, with *'Grace a Dieu'*, and *'Jesus Sauve Tous'*, emblazoned on their fronts.

As we drove out of town along the single-lane rutted national highway, long ago built by US Marines during the 1915-34 Occupation, I remembered meeting Colonel Heinl, one of the officers in charge, in the Pension Tourdot. He would be aghast at the sorry state of his road. I was saddened to see that the good ship Comfort, all white emblazoned with its red cross, had departed. I realized it was a sign that the disaster and medical/surgical side of the relief effort were moving into a new, intermediate phase--life-threatening injuries and severe infections having taken their toll. Because the road was rutted and pockmarked to a depth of 12 inches, it was often filled with sloshing mud, spewed sideways by every passing truck and car.

We drove past all the endless roadside rubble and misshapen carcasses of standing, tilting, and crushed houses that I had seen when I went to Gressier Mobile Clinic yesterday. I recalled again the volunteer doctor who had talked of seeing the sparkling azure blue sea and waving palms, magnificent beyond the ruins and few scattered houses. He had felt it was quite beautiful, until he realized that he was passing a huge continuous graveyard of hidden bodies still entombed by cave-ins on either side. When Google retook its January 23rd pictures for Google Earth Street, the wayside bodies were removed, but many more were still hidden there, never unearthed and recovered.

We got to Gressier in the familiar 1-1/2 hours, and shoved on. The road improved for a while, and my driver began to rocket along, taking no prisoners, careening around curves, even pulling out around tap taps and huge slow moving Camions, barely pulling back in time, playing chicken with daredevils coming the other way. We had had a security meeting the previous night about the abducted volunteers from another organization (I found out that morning they had been returned alive--for a painful price, probably stimulating more greedy thirst for kidnappings). During the meeting they commented in passing that if our drivers scared us, or took chances, we should ask them to slow down. I kept quiet, being used to Haitian driving and wanting to get there fast.

Unwisely, I kept unbuckling my seatbelt to take pictures on either side, at one point being thrown forward as the driver came to a screeching halt in front of a gaping jagged hole zigzagging diagonally across the highway. Usually he had seen what was coming, because other drivers were stopping ahead of us at such hazards warning him. But he was alone on this stretch of highway, surprised at times by these jagged holes, as if Zeus had thrown thunderbolts at the road, leaving their marks.

Finally we crossed the Momance River and my heart began pounding in my chest. We were at Brache, where I always got off my camionette to walk toward the blue Caribbean and my friendly field site. It was vastly different at Brache now—far from just a reflection of fifty years of creeping urban build up because of Leogane's encroachment. When I heard about that guy from my field site in Masson collecting money, I had been afraid, wondering how mud huts and the few old buildings had held up. How had they been affected by the earthquake? My mind's eye had been filled with denial I soon discovered.

When I saw the devastation, I gasped, my eyes welling up, the driver slowing down to see what was wrong. He had no idea I knew any Haitians, especially way out in the boonies. But there I was, a tear-streaked shambles. I let him know everything, about the friends I loved and kept in touch with, and those who were gone or missing, and then he told me what had happened to his family, including losing his aunt and their house. We spent several minutes together on the side of the road talking, put in touch by the earthquake.

I told him I would be visiting my friends at some point, God willing, bringing them photos I had taken of them as children, and of their parents and grandparents, as well as a gift of all my camping stuff and clothes I am using, and (though not mentioned) a financial present. I had American cash with me. In Haiti, the greenback was still king! When I saw USAID tents and tarps everywhere, and the military standing tall and beautiful, I had waves of patriotism and gratitude. Those yellow USAID T-shirts along the way make me smile, as do the Blue of the UN.

To my surprise, as we crested a hill, the dramatic vista of Grand Goave, and Petit Goave on beyond, opened up before us, nestled in a majestic curving bay studded with little harbors. From the pass we could see large ships, and a few scattered sailboats, the harbor cradled by hills and mountains, still fairly green. It was breath-taking, a part of Haiti I had forgotten, since I only saw it briefly on my quick trip out to Les Cayes my last summer there, in '62.

When we entered Petit Goave, my smile faded as I saw the first crumpled buildings, and the myriad clusters of tents, blue (UN) and white (USAID), and other hews of the rainbow from other countries, from Tibet, and Japan, and Mexico, France, and Canada. Petit Goave had been ravaged not only by the 7.0 quake on January 12th centered near Leogane, but by a second 6.1 quake, or massive aftershock, on January 20th centered right on Petit Goave, further devastating the area.

Though more international aid had reached the area than I realized, the situation was still desperate. I was told again that over 920 groups were here right now in Haiti, many trying to penetrate deeper into the southern peninsula despite ravaged roads. Even so, there were vast additional makeshift encampments of motley ramshackle constructions everywhere, made from every salvageable piece of debris found at the side of the road.

Three-quarters of the population were living outside of the houses. That was the incredible short of it, though the long was more complex. People were living outside of their houses because they were destroyed, because they were too dangerous, because they were under reconstruction, AND because they were terrified of aftershocks. In China many villages in quake-prone areas keep deer penned outside the village with a person dedicated to watching them. Deer are uniquely sensitive to the slightest ground tremor indicating approaching danger, great for avoiding predators, AND incipient earthquakes, which always send out pre-shocks as tectonic plates begin slight subterranean grating on each other as they build up for the big one. Well, in Haiti people here are now as sensitive as deer.

We initially pulled up to a gated beachside hotel, the Royal, which looked promising, but then we found out the 'Residence' (actually the Office, where I was to stay), was elsewhere. The Royal rang a bell. That's where I'd be teaching Saturday.

Off we went, finally veering onto a dirt side road off the National Highway filled with small or shattered houses, interspersed with a few larger gated compounds. We finally entering one with a very high wall rimmed with broken glass and barbed wire, protected by a guard. Stephanie, the attractive young director, a French Canadian, greeted me, her cell phone glued to her ear. She was doing a million things, full of energy and excitement. The drive had, indeed, taken just over 2 hours, though I had had my eyes shut for long stretches when we picked up speed—until that wonderful harbor view opened up at the end.

Lunch was underway, quintessentially Haitian, except for an additional veggie stew. Stephanie, our young director, was a vegetarian. And she never seemed to sit down to meals. When and how she ate that veggie stew I never found out. To my surprise, Tom, a tall, imposing Kenyan with a ready smile and deep voice was at table. I had met him at IMC headquarters back in Port-au-Prince. At that stage all I knew was he did some kind of administrative work. Also there was a tall elegant, beautifully dressed black Russian named Jattu. Speaking English, she was a doctor connected with the IMC pharmacy, acting as an assistant to Joanne, our Haitian psychosocial and medical clinic administrator.

Right off the bat, I was confronted with the fact that the nurse clinician had decided not to take the job on my team after all, so a stack of resumes were plopped in front of me. Before I got to that I said I needed to be clear about the psychosocial nurse job description, and, come to think of it, their idea of what my own job description was. Stephanie and Joanne sat down with me and explained that each day I would rotate out to one of the 5 outlying mobile medical clinics and hold a psychosocial clinic with one of the two IMC doctors while the other saw the routine medical patients. I was quite surprised, given the hospital-based plan Lynne had explained to me originally. After mentioning this in passing, I hastened to say that I thought this was the best way to start, allowing me to get to know the doctors where they worked firsthand.

As we finished, Stephanie asked if I wanted to stay right there on the office grounds in a tent out back of the Residence, or back at the Royal Beach Hotel where she had a room reserved for me. There was only one tent out back, but several out front, including Stephanie's. I thought most people were staying out in front. Perverse me, now a tent addict, I chose to stay right there, close to staff action. Nick had thought it a good idea in Petionville. Well, at about 6 o'clock to my surprise everyone disappeared, all heading over to the Royal Hotel, where a nice menu and air conditioning awaited them. And Stephanie disappeared.

Leftover lunch was still on the table beckoning faintly, plastic insect covers frustrating a growing number of eager flies. I deployed myself out back to my tent, at least 4 times the size of my homey Petionville Residence dome pup tent. And it was hot and humid in the still air. I was now down at sea level, instead of up in breezy, relatively cooler Petionville. At least the weather had been perfect since the third day I got here, with beautiful blue skies--no rainy season yet, thank god.

Yet I soon discovered that simply my exertion of going to bed in my tent worked up a sweat, and my morning exercises made me drip. Even so, my limbering up my lower back trumped the humidity, spurring me to stretch. AND, I had lost a LOT of weight, since I could get my knee not just to my chin but almost to my ear. One good thing at least. Why was I losing weight? Not dysentery. It was because lunch often turned out to be dinner, like my first days with Nick. And just leftovers left out for me, the one consumer. At first I hoped Stephanie, who lived in a tent out front, would be joining me. Only problem was, when Stephanie cruised in that evening of my arrival, at 8:30 pm, she blithely informed me she had already eaten in her car. Her veggie stew just sat there forlornly at the table. I had eyed it hungrily but felt it was a no-taste zone belonging to the Director. Stephanie was not exactly a mother hen, especially around food and domestic things. So I ate what I could, including the salad left sitting out all day. I considered it an inoculation to get my immune defenses up, preparing me for whatever lay ahead.

When I looked more closely one day, I was surprised to see something purple in the salad. Beets! They proved irresistible to me, though where the hell did they find beets? When Stephanie stopped gyrating and settled down in a chair in front of her computer, I found out another mobile medical clinic had just been added to my list, this one in Mirogane farther out the peninsula. Also there had been another kidnapping of an NGO volunteer, up at the other end of Haiti, in Cap Haitien far from the earthquake zone. She warned me we were suddenly in tighter lockdown. And I couldn't go out alone. Of course, I had been contemplating going out carousing on the town my first night in deepest quake-torn Haiti, so this was a big disappointment. But in fact I was feeling increasingly land locked. Terra firma was shifting under me.

But something else concerned me more at the moment. I was worrying about preparing for my all-day workshop in two days, and putting their master clinic plan into operation—lots of high profile stuff not good for the psyche of a type A guy supposedly now in retirement. No, abductors weren't my top worry. My life had already been hijacked.

While I was talking to Stephanie in the kitchen, I saw her jump and start to bolt.

"What's up?" I said.

"Didn't you feel that?" she said.

"What"

"That small aftershock!"

Frankly I didn't feel a thing. I had missed my first earthquake. Thinking back, I figured out I must have been in mid-air coming down the step from the bathroom into the dining room. But the next day it was all people could talk about, where they were when it happened, how some people were eating dinner at the hotel and continued to eat, while others leapt up and ran out lest the aftershock continue and the ceiling collapse on them. This was a Pavlovian situation—a bunch of pigeon people, now quite hypervigilant, and over reactive, turned into penned-up post-quake Chinese deer. But then again, if I had been through what they had been through I would be hypersensitive and out the door in a flash. So I cautioned myself to learn, not judge or be a wise guy.

Anyway, I was eager to email Patti, or hopefully Skype her directly, so I went to the Internet, only to find it was down and had been down all day. Stephanie left to do a security briefing and I was alone, electronically isolated. I had no other excuse, so I faced my resistance and began to prepare my talk. Privately, I was feeling a bit lonely, cut off from Patti, no real food, no company, and a strange new tent to sleep in.

Well, I frittered around for about an hour, reviewing things, then prepared to hit the hay, or, rather, the air mattress, finding that I was swamped, disorganized and bushed. Where was my psychiatric sidekick, Peter, when I needed him? We had all been given Haitian Digicel cell phones with our own personal number. All Haiti seemed to have one. I gave Peter a buzz, told him I'd lined up four nurses for us to interview when he made his trip out to check up on me the next day. "Great, Kent, see you soon." Click. Peter wasn't one to gab or waste motion. I was alone again. Haiti and aloneness were plaguing me again. Was my old second summer nightmare coming back? How can you feel alone in a country where people are virtually living in each other's laps? I was amazed how umbilically disabled I felt with no Internet.

But sleep beckoned as I slithered into my sleep sack and bounced onto my air mattress--except that my friendly night roosters were now replaced by the late night carousing and radios from some nearby Haitians, and NGOs. Turns out we were next to a local radio station, which played what was on the air over a loud speaker out back so family and friends could party a little. But, I was so exhausted all was somehow music to my tired ears. I found myself ruminating about the mission I was on for all my Haitian friends, and consoled myself with the secret thought that I would only have to be in this altruistic humanitarian mess a little over three more weeks. I could stand it, or could I? I'd give it my best shot. Only my powder wasn't dry because I was sweating so much. Well, my chance had come.

Saturday, March 13: First Teaching Seminar: Traumatology 101

Saturday came around fast, and I had to stand up in front of all the local Haitian

IMC clinic doctors and nurses, people I had never met, and talk for 5 hours about the psychosocial impact of what they and their patients had been through already themselves, something that daunted and humbled me (when I first typed this into my diary, I wrote "dumbled and haunted"!). Of course, this was also their first look at me. For some reason, I had trouble falling asleep that night, waking up again in the middle of the night. I spent from 4 to 5 am floating on my air mattress, dealing with my anxiety, finally pulling myself up by my bootstraps. I found myself remembering that private voodoo session Ternvil had given me gratis 50 years ago in which Maitress Erzulie Grand Freda reassured me that if I worked hard, I would be successful and maybe come back to help the Haitian people with my new skills. It was cuckoo of me to be anxious about what was now my strong suit. Somehow the gist and the words for my talk took shape deep within me, and, inspired by my Haitian friends, their strength in disaster, and the mission of IMC, I took heart, found my courage, and fell back to sleep, until 5:45, when the cock crowed for me.

I must admit I felt encouraged by the great IMC staff--Stephanie, Joanne, and Jutta, and especially Nick and Peter, all having shown me the way. And I had copious lecture notes form that wonderful English child psychiatrist in charge of program, Lynne Jones. Joanne had all our handouts with her as we drove to the Royal Beach Hotel for my workshop, with sections entitled, 'Mass Trauma, Loss, Grieving, Front Line Mental Health work, Treatment, Triage, Symptoms, Major Mental Illness.

PTSD, though popular with the press and the world, would be only a small part of my seminar, since, surprisingly, statistically it is a small percentage of conditions actually diagnosed in mass disasters, provided aid and care are given correctly right away. We also would cover pre-existing epileptics, manics (bipolar cases), severe depressives, and chronic schizophrenics who were stressed, lost their medicine and/or their treating psychiatrist because of the quake, as well as new onset, or 'first break' cases caused by the tremors. When and how to use scarce psychiatry and inpatient hospitalization would be emphasized. We hoped as our IMC mission objective to leave a legacy of psychosocially competent front line Haitian family practitioners when IMC left in two years.

Despite my nightmares (or wish?) that no one would come, there were 12 Haitian doctors from our 5 clinics, and 19 nursing, plus some people from Notre Dame Hospital. We had done some liaison work with the Croix Rouge, who helped with the hospital staff. The Red Cross group sent 5 residents and nurses. Peter accompanied me to help introduce and launch me properly. It went fairly well. I even spoke a little Creole to the group when the translator had trouble with my medical English and concepts. I did a group exercise teaching them relaxation and imagery techniques useful for interrupting cycles of anxiety and repetitive anxious and depressive thoughts. While in the relaxed state, I had them visualize where they were when the earthquake struck, helping them recapture and work on their own inner experience, to increase their emotional availability to each other and their patients. I emphasized cost-effective front line stress reduction nursing group sessions for people they could identify and recruit. Just one or two sessions would be effective, serving more people than one-to-one work, so I was modeling this approach with them.

I then had them pair up and tell each other about what they had been through, and then asked them about it so we could use their own range of normal expectable emotional and mental experience to illustrate the symptoms and natural range and phases of recovery, establishing the wide scope of normal before one even got into making true diagnoses. You have to know normal to see where abnormal begins, and you have to be flexible and wise in your judgment calls—especially when the numbers are vast and the resources short. We talked of acute stress symptoms, and how rare chronic serious post-traumatic stress disorder is, and that early intervention on the front line could reduce the occurrence of PTSD dramatically. Among other things, I mentioned that rape, abuse, and violence, as well as the trauma of war and disasters, can cause PTSD, particularly when doctors as well as personnel and civilians see and deal with mutilation or dismembered bodies. My talk was geared to war zones and earthquake areas like Haiti. I heard that even people just looking at the stacked bodies roadside on Google Street had recurrent nightmares.

Once again, like Mars and Kline, the great turnout was assured by a delicious Haitian buffet, with red sauce chicken, a conch stew, tasty vegetables, and fruit galore. At least I could eat well once a week. In her evaluation, a great nurse I came to know well because of her sense of humor, skipped all the post-lecture written questions and went right to her bottom line: "there was no pie or cake for dessert" was all she wrote. And I should point out that she was already the heavyweight of the nursing group.

During the lunch break, as everyone else made a mad dash to be first in the buffet line, one young doctor came up to me and said he was unsure about the purpose and usefulness of the imagery recall exercise. When I explained it again, his eyes rimmed with tears. He told me about pulling children, some dead, some gravely injured, from under crumbled concrete slabs in the house next to his after he and his kids managed to get out safely, just before his house finally collapsed completely. We talked at length, he was grateful, and I thanked him for having the courage to talk with me. I knew I'd be working with him soon in one of my weekly clinic rotations. I was deeply moved by this experience, and by the group at the conference. It certainly broke the ice for me, and I hope for them. I felt poised, ready to go out to meet them, work along side them, and treat their patients with them. All readiness and no experience yet. How would I do? I had a strange thought, *Erzulie, please come up from the abyss. Be my goddess and protect me from demons. Help me to find the healing words I need.*

In the afternoon I stressed the importance of their health care presence at the clinics, which were strategically placed in or near the tent camps and the destroyed villages, reinforcing with immediacy the impact of their caring presence and activities, their laying on of hands, their quick but careful exams, including their mental health first aid and triage. As front line workers, their work and reassurance gave hope and momentum to recovery for this vast impacted impoverished, yet strong resilient Haitian people, helping them move along in their expectable stages of recovery from a mass disaster. I emphasized that they should think of how to normalize their patients' 'abnormal' experiences, maybe based on their own seminar guided-imagery experience. This might help correct their own and their patients' dire fearful self-diagnoses of their symptoms, their weird thoughts and feelings. They needed to be careful not to overly 'pathologize' what they saw, by coming to know the normal stages of mass disaster recovery, to help their patients avoid getting stuck and becoming chronically symptomatic.

I emphasized the individual and group importance of their presence at their Clinics, that they were, just by their presence, and their laying on of hands and manner of caring, a transference object of great importance for the tent camp and village they were in, becoming a healing beacon in the troubled mental sea. Although I knew they felt guilty about the long lines every day, and their brief problem-focused encounter with each patient, I said that their Haitian patients were used to waiting for care, care like they had never had before, and that even waiting in the clinic near their doctors was curative as part of the placebo effect. Just knowing the Clinic was there mentally, and spreading the word to the camp or village, was good for people. I repeatedly emphasized that they were on stage for their patients, as they delivered care in these open tents with every one watching, and that they should never underestimate the importance of using themselves as a powerful part of the healing. Patients who sit and wait are in a healing presence that sets the mental stage and also cures in its own right.

IMC stressed how much shelter, food, water, security and reunification or 'retribalization' means (including connecting people with religious and secular groups if their own families are shattered or dead) in providing the substance and holding context for their recovery from mass disaster. Several nurses and doctors shared their own family experiences, and then, feeling safer and more encouraged, their patient experiences. Then I looked directly at them and asked them how they themselves were doing, asking for a show of hands about how many had lost family members, how many had their houses destroyed, how many were living in tents outside their houses, and how many in tents in the camps.

Though maybe more fortunate than some, all were traumatized, and about a third had been deeply affected one way or another, many with significant losses, several now living in the camps. When the subject of tents came up, I noticed two nurses looking down and huddling privately. I finally asked if they could share what was going on. With some embarrassment but plucky honesty, one nurse confessed she didn't even have a tent yet and was living outside with family members in one of the camps, grateful the rains hadn't come, and proud she made it to the clinic every day to work, somehow looking clean and kempt. It seems tents are now in short supply in Haiti and there is still great need. The clinic was clearly a beacon of hope, care and support for her, too. And her own experience gave depth and meaning to her work. Anyway, I asked my group after the tent revelation a telling question: "Who takes care of the care takers?" I talked about the importance of self-care for long distance healthcare workers running in a mass marathon. We chewed on this question for a while, and I, among others, suggested things they could do to care for themselves.

Right after the Seminar, before Peter took off for Port-au-Prince, we reviewed the participants' critiques of the seminar and then did interviews of perspective psychosocial nurses for my team. Peter zipped through the critique of my seminar work, smiled, and handed them to me. Most were moderately positive, but a few were fairly negative. This either meant I was trying to teach too much, flooding them, or I was incomprehensible or losing something in translation. A few said Tessier was too soft spoken and couldn't hear his translations. A number mentioned they wanted more handouts from me, and in French or Creole. Many wanted more active participatory teaching exercises. I read it slowly, took it all in, and tried to keep up my smile. But privately I felt a little crestfallen. I would have to do better. Peter said I should take it with a grain of salt, but give them a big dose of activities next time, and throw lots of paper at them. "You know, give them a big taste of their own critique medicine."

Peter was seasoned, smart, organized, and quick about everything. For the nurse applicants, he had a set interview, clarifying and verifying the candidate's CV, asking about all their psych training, and then asking them to describe in detail an interesting patient they had worked with closely. To our surprise Mars and Kline and its good teaching kept coming up. Maybe it wasn't the snake pit back then that it now seemed to be. Next Peter liked to challenge candidates, throwing out a hypothetical situation and asking how they would handle it. Some flubbed this; others were dull and inarticulate, lacking a sense of personal rapport. Some were solid and passable, and our last was very bright and psychologically minded. She also happened to be quite well dressed and attractive. Her name was Nathalie. But she lived in Port-au-Prince. When we asked about her willingness to work in Petit Goave, she said she was used to traveling and living in different places, and would be happy to come to Petit Goave for a while, but would prefer to work in Port-au-Prince. She rejoined the group outside, waiting for our decision.

Hands down, Nathalie was the one for both of us. Then the atmosphere seemed to change, Peter talking vaguely about maybe we should wait on choosing a nurse for my team, saying, "Kent, maybe you ought to get settled into your routine, before taking on a nurse." I had the distinct impression that Peter wanted Nathalie for something in Port-au-Prince. I was a little confused and dismayed, but felt there was some merit to my settling in before taking on a nurse. I really felt I needed to get my feet firmly on the ground.

With the interviews completed, I found I was still thinking about my seminar and especially the nurse without a tent. Now it so happens I had a tent with me that I might not need to use, given the Shelter Box I was in. We seemed to have enough tents in our privileged IMC 'camp'. Though it might not sound like it at moments, we were pretty well taken care of, mostly. My extra tent was given to me by a fine, taciturn friend, Norbert, the Swiss German husband of dear Kim, our American caretaker for a set of ancient rooms we own in an Italian hill town called Fanghetto. When we came for my last supper before Haiti, Norbert was quietly moved by my trip, disappeared for a moment to his basement, and emerged bearing many gifts. He gave me a lot of camping equipment, urging me to give it away in Haiti when I finished my work, bless his soul. So I planned to give it to this nurse when I went to her clinic, but secretly, so as not to stir up envy, or a perception of favoritism by IMC. But I thought the other nurse huddling with her might also be in the same plight, so I wanted to seek a more systematic solution, asking wonderful Stephanie if she could get tents for her/our own people, at least two. I didn't want to wait on bureaucracy, but Stephanie seemed to make glacial administrative mountains melt for her people. She was my kind of NGO person!

My only complaint about her was that she kept giving me serious security grief about letting me go to my old field site, saying I couldn't go right now, or in the foreseeable future because of the increasing abductions. She went on about how one abduction can cost an NGO $30,00 to $50,000, which would mean that hundreds of refugees would have to go without food and care because of the lost funds. There was something strange about the way she tightened up telling me this, and how guilty she sounded--specially her using 'refugees' that way. No one was calling the displaced Haitians 'refugees'. So I asked Peter and found out Stephanie herself had been abducted for 25 terrifying days in Darfur, before touch-and-go negotiations got them to release her. But all she seemed to feel about that, besides wanting to protect us, was terrible guilt about how much she had cost IMC for her ransom. She seemed preoccupied with the lost money depriving the people in Darfur she loved so deeply. The more I thought about it, she gave me a new sense of the concept 'survivor guilt'. She had a lovely amulet she wore every day, given to her in gratitude by her people in Darfur. I found myself tearing up again as I wrote this, feeling a mix of awe and admiration for this woman. And bless her soul, Stephanie said she felt she could probably get some tents for my nurses in the next week or so.

Speaking of Stephanie and tents, when I was making a preliminary tour of all the mobile clinics on Thursday and Friday, I visited an IMC clinic called Beatrice, at the top of a high hill, serving a tent city around a small quake-ravaged village, and drawing from far into the mountains above.

The next day she had an emergency call from Beatrice. A big wind had come, whipping through the tent city and blowing our big Beatrice Mobil Clinic tent down--with everyone in it. As usual, they called Stephanie to rescue the situation. Luckily, no one was hurt, just a lot of ruffled feathers. Stephanie spent the afternoon recruiting willing town folk to help put the Clinic tent back up fast. They loved the Clinic and pitched in with vigor, even though the local government got their feathers ruffled too, saying she should have gone through them for the labor. When we were at Beatrice, the Clinic had lots of patients, and Joanne and Jutta came through while I was working with the Haitian doctor, doing a quick administrative review and update of supplies and equipment. We also visited Petit Guinee, a destitute and now ruined seaside village. And finally, Trois Soeur Clinic, in a tent city around a convent.

After my Saturday Seminar, I felt pleased (and relieved) enough to treat myself to a Haitian lobster dinner at the Royal, after two Barbancourt rum and cokes, later listening to the 'Strict Badou', a jazzy, professional Haitian group, with appropriately strong island flavors. And then I slept very well finally. Even the roosters failed to rouse me. The next day, Sunday, would be a day of rest, and a chance to write. I made an omelet for Stephanie and me, and gave my novel, **Body Sharing**, to Peter who was complaining about nothing to read. That was all, except for preparing my materials for my first clinic tomorrow. I looked forward to beginning my clinical teaching and training, though I've already begun to worry a little about my lecture in Port-au-Prince at the Mars and Kline Psychiatric Hospital next Saturday, possibly on Adolescence. I had the material to give a lecture on Eating Disorders, replete with slides, but somehow that didn't feel quite right for down here. Anyway, my inner Type A personality was already possessing me again, interfering with my day off. As I said, character is destiny. I was already possessed by Type A when I was in Haiti 50 years ago, but Ternvil, my Voodoo priest, had never heard of that particular god.

Monday, 15 March: Dreams and Nightmares

I was in a narrow muddy rutted road between tents when I saw a green steamroller, or maybe one of those big, bug-like French street sweepers coming straight at me. Nowhere to hide. So I quickly rolled over to the side in a panic, landing in a jarring heap. I awoke, discovered it was 4 am and that I had landed in my suitcase. As my mind locked in, I realized I was facing my first clinic in Petite Guinee, a beautiful spot on the edge of the azure Caribbean in perhaps one of the most impoverished destroyed areas in the Petit Goave region.

The previous day I had met Chrisie and Laurie, a doctor and nurse from Johns Hopkins, fresh from Port-au-Prince and eager to take over the clinic from departing volunteers. From my visits to other clinics, I realized I was ill prepared logistically, even for the basics. My mind started grinding out my need for 5 folding chairs (me, my Haitian general practitioner (Dr. Charline Louisjas), my trusty interpreter, Tessier, and the patient (plus his mother, if a child), a folding work table, my pharmacy supplies, and chart materials, plus lots of water. It would be sweltering. I had been promised a little tent for my clinic and I realized nothing had arrived. No one had mentioned anything about this. I guess I was going to find out what it was like to be running a mental health clinic tentless. Making mental notes, I fell back to sleep until 6.

My Petit Guinee Clinic was supposed to start at 10 am, two hours after the Guinee staff picked up their meds to go out, set up, and get started. So I had arranged to meet Tessier and our driver at 9. Only that morning things went haywire because of transportation SNAFU's. I had to pull up stakes at 7:30, loose ends trailing. If I hadn't been sleeping in a tent right there on the Office grounds I wouldn't have been able to deal with all this. The pharmacy was at the Office and all transportation launched from there by a great group of drivers.

Turns out there was method to my tent madness after all. I was strategically located in the thick of things. Serendipity came out of this chaos. I discovered I liked going out with the teams at first light. It was fun chatting with the drivers, medical staff, and eager new volunteers, who were amazed at the extent of the damage so far down the peninsula. One quoted a most recent CNN commentary, which said the rubble from the Haiti earthquake would fill the entire Washington Mall to the height of the Washington Monument. My heart caught in my throat as a realized the fresh impact of what they were seeing. We saw a house totally destroyed, with a slanting slab of roof, now taken over by goats standing at the peak. At least they wouldn't be eaten at night, unlike the 'free-range' chickens with nowhere to hide. I now had more sympathy for the roosters and realized why they were crowing at all hours. Packs of hungry dogs roamed the night, seeking whatever they could scavenge, given the scarcity of food and leftovers.

On the way to Petit Guinee we drove through the poorest section of Petit Goave, and, though in a beautiful seaside location, also one of the hardest hit. I was privileged to see how the staff set up the clinic. Patients were already sitting on fractured cinder blocks for stools, squatting in classic 'Haitian style' all around the periphery, perhaps 75 people strong. At Petit Guinee Clinic, some mothers were breast-feeding, other mothers and fathers holding sleeping children, all eager but respectfully waiting for a turn. All huddled under a huge, slightly twisted corrugated roof with open sides. Tables were set up, and blankets suspended and tied into makeshift walls, giving a semblance of rooms and privacy. Chairs were at a premium, as were tables, so my nightmare wasn't in vein. We came well prepared, all very useful as the clinic began to roll. I was given a corner up on a cement dais, a remnant of some sort of stage with a shiny pole in the center and an old bandstand.

Dr. Affricot was chief of the clinic that day. So there I was on stage for my first teaching clinic. Without a drum roll we saw our first patient. But what kind of place were we working in? The remains of murals covered one wall, but I paid no attention. We had things to do. My 'office' was on a cement dais, a long table cordoned off by army blankets. As I walked up, one of the volunteer doctors was taking a blood pressure as a Haitian doctor watched. They stepped away, and our first patient came in.

Pierre, a shy, taciturn eleven-year-old, was sleepless, constantly hearing the cries of a baby, and the voices of dead neighbors. He had been holding a neighbor's baby when his house collapsed on him. His mother could only see his head when she tried to rescue him. He tried to protect the baby in his arms, but it was gasping when they got him out. The baby died on the way to the hospital, crushed in his arms. He felt horrendously guilty, not helped by the baby's angry grieving parents, whose house had also collapsed. His mother added they weren't really talking about him personally, but he felt guilty, even for surviving. He had had a friend die 3 years earlier and heard his voice for a long time, thinking at times he even saw him in groups of children, until taking a second careful look.

I worked with the doctor to do the interview, using the interpreter to get the story details and give feedback, providing guidance, at times even speaking in my rusty Creole to the boy and his mother. He had made it through the mourning of his previous friend. So he could do it again. We told him he had more complicated grief work to do this time, but by past experience had what it took to work his way through this one too. We said we felt he would do fine. We told him and his mother he was doing too much emotional work at night, giving him bad dreams and sleepless dreamy voices during the day. I explained they needed to bring this into the daylight when he could do more effective emotional work.

The mother was advised to have a little session with him each evening to gather all his worries into her mind and arms, helping him clean and clear his mind, reassuring him she would work on them for him so he could sleep—kind of like what you do with Guatemalan worry dolls. She should also tell him he had done all he could for the baby. Nobody could have protected him more, not even his parents. Because he was a shy boy with a strong conscience, making him very self-critical, she needed to tell him to ease up on himself. We gave no meds, but rather a follow-up, saying we felt they would be a good team during their healthy homework. They left encouraged and armed with active self-help they could carry with them

The next woman had severe palpitations. She was on the way home when the earthquake hit, seeing friends in front of their destroyed houses wailing for dead or missing children. She rushed to see how her 5 children had done, finding 4 alive in front of their collapsed home, 'Grace a Dieu'. But her 5th had not made it home from the school, which had partially collapsed. She wanted to rush out to find her, but her children reassured her she would come home. And she finally did, with stories of other kids being hurt or trapped. It was 3 days later that her short agonizing vigil waiting for her daughter triggered severe palpitations.

She had pre-existing high blood pressure, was on a medication, and worried her heart was giving out, with bursts of rapid beating (palpitations) making her feel she was dying. She let us know she was helping many of her grieving friends, and felt her heart problem was physical. But she had never had this before, except slightly walking up steep hills.

It became clear, after taking her blood pressure and taking her pulse rate, and listening to her heart, that she was physically okay, though we agreed she should see her doctor to get checked out, maybe even have an electrocardiogram. But I told her we knew what was going on, and that she had the strength and intelligence to work this out, letting time and simple techniques restore her trust in her body and in life. I explained the endocrine fear response and her tendency to make scary self-diagnoses escalating her panic. We noted her previous fast walking would make anyone's heart beat faster, and that the new bursts of heartbeats were different, a normal fear response that had gotten a little stuck. To deal with that she needed a couple of techniques to counter thoughts or noises, or after shocks when they triggered them. We taught her to blow into a sack and the Valsalva maneuver, like when you bear down to grunt at toilet, or during childbirth.

The Valsalva causes a neurological reflex (vasovagal) that slows the heart. You may have read about this in diving mammals. We told her to use it to interrupt the beginning palpitations. This both works and is a cognitive distraction. Just knowing you could take control helps a lot. We also showed her the paper bag sealed around the mouth re-breathing technique used for hyperventilators, but also for palpitations. The 'sack' re-breathing technique decreases O_2 and increases CO_2, and distracts—decreasing her overbreathing and tingling and dizziness caused by excessive breathing and increased O_2 levels. We also urged her to be a smart scientist, noting down obsessively each time she had such an attack, so she could outfox the triggers, and disconnect them with an 'I told you so", just as she could help her friend do. She needed to be a kind doctor, and not scare herself. She got the hang of it, and understood the psychology and physiology of it. She was a schoolteacher so I suggested she could help teach this to scared symptomatic friends, as she herself got good at it.

Another patient was glassy-eyed and depressed, showing us a certificate of scholastic accomplishment earned by her 21 year old son. His handsome picture smiled up at us from the certificate. Between sobs she told how he was teaching in Gressier, away from home for a while, and was crushed in his little room by the earthquake near his school. She was consolable but in deep prolonged, but not arrested mourning. Yet it bordered on depression. We listened with near reverent attention, checked on her friendship and religious network, and noted she had high blood pressure. She also had serious insomnia. I suggested they add Atenolol, both a relaxing sleep promoting and anti-hypertensive medication, hoping to help her through this sad, sad passing. She had other children to live for, but we would follow her up closely next week just to make sure.

We saw other patients today, and as time went on I relied on the Haitian doctor more, since we were hoping to give our clinic doctors increased front line mental health competence, a good sense of basic psychotropic meds, and diagnostic acumen for triage---and referral, if absolutely necessary. But where could we refer? The psychiatric hospitals were mostly destroyed or seriously overcrowded and understaffed. There wasn't much psychiatric care to go around, and most people, even if deeply affected, were, with simple help, resilient and self-righting—if they had their basic needs met, that is, shelter, water, food, and security, plus some social connectedness.

We had one other woman in the Clinic today who lost one child, an aunt, and her house. Her business establishment also collapsed and then was looted, and, her van was trapped under a concrete wall. So she and her family were without even a tent and no means of livelihood, after enjoying a comfortable, productive middle class lifestyle. She was depressed, and, I sensed, quite angry underneath. Because of this and a sense of pride, she was unable to reconnect with, and in fact avoiding, her Pentecostal Church. She seemed close to needing anti-depressant medication, but we gave her a light sleep med at this stage. She and her IMC Haitian doctor preferred it this way. He pointed out to me one had to be on costly antidepressants a long time, and we sensed she might come around the corner if we waited. We planned to see her again next week, just to make sure. Taking the mental pulse and providing close follow-up were the key. We didn't want her remaining children to suffer a maternal suicide because we were too conservative and cost-conscious, given everything else. But for her this proved a wise choice.

When I got back to the Residence office, I had to go with Stephanie and Peter to the Tuesday meeting of a Regional Psychosocial Cluster for all mental health leadership, which was boring for me at this stage, but also a key brief encounter with the hospital administrator, since we wanted to get approval for us to be a hospital resource for medical and surgical inpatients having enough of a problem that psychiatric intervention and meds are indicated. Though I loved this kind of hospital consultation-liaison work, I was already feeling spread a little thin. I also heard there was an administrative movement afoot to base me at the Notre Dame Hospital for efficiency's sake, reverting to Lynne's original plan, instead of continuing to go out to the individual clinics. I felt this was a bad idea for both the Haitian Docs and Nurses, and for the patients. As I got whiffs of this I began lobbying lightly against it. We would see.

When I was hanging around after dinner, everyone else off to the Hotel, I noticed Stephanie was sitting in the office in an especially good mood. I didn't have a clue why. She seemed to be smiling at her screen saver, so I took a more careful look at it and remembered something she had said to Jutta. There on the screen was her friend in Darfur, for whom she had been keeping a vigil. They had worked together and he had been abducted for ransom. This had happened shortly after her abductors had released her back in Darfur—something she had not yet told me about personally. That was way back in November and it was now March. After all this time, and her losing all hope for him, she told me she had just found out from him he had been released. She was sitting there thanking her lucky stars. It had been a long hard vigil. I didn't interrupt her reverie after that, but felt incredibly glad for her, and her friend.

Stephanie was a vegetarian, not really interested in food. She tended to eat at sporadic times. For myself, (and her if she wished), I took matters into my own hands and cooked a vegetarian lunch Sunday, and also made dinner that night, linguini with olives, onions, tomato and PREGO mushroom spaghetti sauce. I offered some to her, but she wasn't interested—until savory smells wafted over her direction. Then she suddenly joined me. Yum, after my fashion, and Stephanie liked it.

While I was cooking I overheard Stephanie talking about leasing a new group residence for us, with our move date depending on the installation of barbed wire atop the high walls. I had mixed feelings about this news, since I liked living behind the office in a tent, and didn't want to give it up, kind of like the film *Woman of the Dunes*. Plus I would not be right there in the thick of things when the docs came for their meds and our group departure. But I gathered the move was inevitable, and I needed to get with the program. That seemed to be my new theme song. Compared to my footloose anthropology days, I was becoming a company man. Sigh.

Knowing how many things Stephanie was juggling and her disinterest in food, I suspected she hadn't told our cook. So I called Crystal over and clued her in on all this, helping her add up the count for all staff and volunteers she would have to cook for. The total would be 7 for dinner each night in our new Residence setting. I explained that the volunteers were used to hotel choices and lots of food, all coming from a Port-au-Prince hotel, so she should plan on ample food for hungry mouths. I must admit I was protecting myself from the starvation of miscalculation. Her eyes got wide, and she thanked me, shaking her head. My Creole wasn't too bad in a pinch. Anyway, I had the fantasy I was going from a minimalist tent to a more sumptuous situation.

From a material and creature comfort point of view I already sensed I would be using Haiti and this experience as my litmus for checking out how materialistic I had become. This has, at moments, been a physically depriving and materially sparse experience, yet nothing compared to the Haitians.

As I did my diary every few nights, I discovered writing about my cases was truly 'bibliotherapy' for me, which became part of my self-care, something we emphasized around here. It helped keep my head above water, so I didn't sink into a teary abyss. When we opened our hearts and our minds to the suffering here in Haiti, we rapidly discovered some of her people were drowning in sorrow, clutching us doctors and nurses like someone drowning, threatening to pull us all under with them. I don't want my day job to lead to accumulating nightmares that might wear me down.

Monday, March 15: Close Encounters

Monday night I had a spirited discussion with Stephanie sparked by her mentioning we'd be moving my beloved mobile mental health clinics from their current countryside sites back into the Notre Dame Hospital right in the middle of Petit Goave. To her surprise, I went off like a sky rocket, ranting about how these poor patients were already stressed to the limit, didn't have any money, and found it hard just to come to our nearby clinic, and, their Haitian doctors and nurses needed teaching where they were, and not be forced to travel into town to see me. Plus, we shrinks needed to get a first hand sense of what they were up against out there in the tent camps. This is what community psychiatry is all about.

Faced with this unexpected passionate defense of our mobile clinics, she was taken aback. "Why are you so angry at me, Kent? We're professionals discussing alternative plans."

"I just feel strongly, and want to express myself forcefully. I'm not angry at you personally."

"Well, that's not how you sound. And remember, I'm just the administrator, and YOU were the one who told me the new psychosocial clinic wasn't supposed to be out attached to the mobile medical clinics, according to your boss, Lynne, and her master plan. Now you're telling me the opposite. What's going on here?"

"You're right. But I'm glad we did it the other way. It's best. And my experience proves it."

Stephanie got very quiet, and finally said, "I probably shouldn't say this, but I have an idea why you're so heated up and don't want to lose going to the mobile clinics. You have personal interests, underneath, so maybe as a volunteer you're sort of a tourist, wanting to see these exotic places."

It was my turn to get quiet. Her comment hurt. "You have a point," I finally said, "but I'm no accidental tourist. I'm using everything I know for our Haitian patients and their families. You're incredibly dedicated and hard working, like Lynne, Peter, and Nick, but I think you and Lynne have lost sight of our mission. Even in Port-au-Prince IMC teaches IN the mobile clinics, not requiring just the doctors to come to the hospital, even though Nick and Peter also see hospital patients. I feel that's important, too. I spent 5 years of my life doing just that kind of work. I made the rounds with Nick and Peter in Port-au-Prince, so know what I'm talking about, and feel it's best."

I knew Stephanie was working like mad, and was perhaps the most dedicated person I had seen yet. With Lynne, Peter and Nick, and all the IMC team I had met quite a few. And, in fact, I agreed with her about the crucial objective of also establishing a secure and effective psychiatric base in the hospital, and was already going about doing that. But to move out of the mobile clinic sites seemed wrong."

Stephanie countered, "But still, you're so angry at me, Kent, why?"

I replied, "Well, maybe at Lynne but expressed toward you. Back home my wife would call this just a good healthy discussion. You should see me when I really get mad." She shook her head in exasperation. We parted agreeing to disagree.

The next morning Stephanie was all smiles but I felt the ice. I really missed coffee with Patti that morning. I'm not a big coffee drinker, but love making hers, and frothing that milk just like Rob Low, the Coffee King, taught me. I was missing her, but would just have to wait. Oh, and Patti and I were good at making up, too.

For my 'safety', Stephanie told me something very interesting, just to make me feel at home in my tent behind the office. Now, every night, before turning in I hunt for the tarantula she saw the other evening back behind my tent. She knew how to give a guy peace of mind. No sightings yet. Tarantulas didn't really bother me any more, or so I thought.

When I was in Haiti that second summer long ago, sleeping on my cot with the mosquito netting draped over me, suspended from the ceiling like a gossamer teepee, I often heard a scratching under my cot as the cool of the night entered my dirt-floored room. Finally, one evening when the racket irritated me more than usual, I turned my flashlight toward the trunk bottom next to my cot, and thought I saw something big coming out of a hole. But suddenly my view was obscured by something hanging on my mosquito netting right in front of my nose. When I focused on it a huge tarantula was climbing up the netting right in front of me. I freaked out, screamed, and Joselia and Ternvil came running from the next room to see what was wrong. They chased the thing away, plugged up the hole with a rag, and calmed me down.

The next day I was typing my anthropology field notes by the window. From the corner of my eye I saw something reaching its hairy black legs onto my left shoulder. I still have flashbacks from that moment. As you know, one of my flashbacks just happened again flying down to Haiti on Air France. That original tarantula was apparently coming back home via the window. Usually an animal lover, I took revenge by pouring an entire bottle of expensive 5-star Barbancourt Rhum down his house hole and lit it, much to the laughter of my Voodoo Priest, Ternvil, and his wife Joselia.

Bonheur, my buddy, the top voodoo drummer, happened to be walking by at that moment, and ran over. When he heard what was going on, he cracked up. He found the tarantula that had climbed on my shoulder, picked it up with his bare hands, put it on his arm and let it crawl slowly up his shoulder, over his head, and down his other arm, chuckling the whole time.

Then he told me a little secret. "They have a nick name in Haiti for tarantulas—'Nothing Crabs', meaning they look scary but can't hurt you, only a little nip if you pester them, and occasional allergic reactions."

"Okay," I said, "but you can still take your pet home with you, Thanks anyway!"

I was pretty well convinced they were harmless by then. But I had been psychologically traumatized, and no one from IMC was around to help me out back behind the IMC Office. So I was doomed. Some of you know that I started my novel, **Body Sharing**, with a Haitian tarantula torture scene. I wonder why? The same reason Stephen King has a recurrent image in most of his novels. As a young boy, King saw the remains of a neighbor kid struck by a train above his house on an embankment, and if you read his novels, most of them have a frightening scene involving a bright light bearing down on someone, finally smashing him to smithereens. Not always, but sometimes, trauma has a way of etching itself into your brain and coming out in weird ways.

But enough reminiscing about times past. I turned out the Pharmacy light, crawled into my sleeping sack, and drifted off—'to sleep, per chance to dream': About *being lost in an apartment complex, late for a meeting, ending up taking a final exam which completely baffled me. I looked around the room noticing lots of bright young people plowing through it with ease, finishing early. One face stood out because of his knowing smile. It was Hugh Hill (the younger), clearly acing the exam. I decide to take the test another day, get on with my next appointment, a little disappointed with myself, feeling I didn't have all the answers anymore.*

At least the dream didn't wake me up. Until I had to pee. Now who wanted to go outside the tent in the dead of night when tarantulas were lurking about, plus waking up too early would be tiring, given my upcoming teaching clinic. So I brought along a handy little bottle I appropriated from my Paris hospitalization a year ago for a septic elbow. The only thing is, I hardly had to use it because of dehydration, despite keeping hydrated by steady drinking. What really amazed me was that during the day at these mobile teaching clinics, I drank steadily, but could go 8 or 9 hours without peeing, something my wife will tell you is unheard of in Paris for this older man. I had the thought, *Bye, Bye, FlowMax, you're history now.* My whole water metabolism had shifted. Under the corrugated clinic roofs or tents we saw patients in, it constantly felt like an Indian sweat lodge.

Despite such advanced nocturnal planning, I was too awake, though in a twilight state, listening to rooster calls sweeping over the countryside like amber waves of feathered grain—until my local rooster came home to roost. He was so loud that sleep was impossible. Then a distant foxhunt began with dogs baying, apparently running in packs around the neighborhood, waking others as they ran. Occasionally they would bite each other, no fox in sight, and you could hear the victim go yelping off into the night. It was quite a ruckus for a tired old man—except I was lit. Weight loss and maybe Chloroquine light my fire. *'Yon Casius has a lean and hungry look'* ran through my mind. Yes, I was taking Chloroquine again. I had planned to take Proguanil this time for my malaria and parasite suppressant. Mosquitoes, and what they spit into my blood stream, bugged me. Sometimes paranoia was useful. Fifty years ago they also gave me that Cocksackie B virus and Dengue Fever (my 'heart attack' and 'breaking bones'). If you take a suppressant whatever parasites those mosquitoes spit into your blood got killed. I guess in my dotage I was feeling more like a sitting duck, or more, a juicy pin cushion rather than an invulnerable 20-year-old. Unfortunately, the Proguanil sent my blood thinner, Coumadin, sky high, suddenly making my blood too thin, risking bleeding. So back I went onto Chloroquine. I just hoped my tinnitus--that ringing in my ears--didn't get louder. When I was lying awake, if I thought about it too much, my tinnitus actually did seem to be getting louder--but probably no different than usual--just the hysteria of nocturnal ruminations. Luckily I was a retired shrink now, so I didn't really have to listen to anyone carefully any more, especially my own craziness. Of course, there were exceptions. My wife Patti. And all my new Haitian patients. I had to keep attuned.

As my early morning clock ticked away, my more rational thinking locked in. I realized I was going to the Miragoane Clinic today, a newly started clinic that Peter hadn't checked out yet. Afterward I would be going to the Notre Dame Hospital to 'show the IMC flag', create more visibility, and drum up some business. I also had heard a private lab across the street might be able to do a coagulation time and INR, since I've been flying blind on whether my blood was too thin. Back in Paris I hadn't had time to check my blood again after switching back onto Chloroquine. My little white lie to Patti that all was under good control had become a worrisome reality for me now. Was I at risk for a bleed? I wasn't sure.

Yep, it was self-care day for me, before I went to see if Notre Dame's Haitian residents might like some teaching. Several had come to my Saturday lecture, as you know. Maybe they would nibble at a seminar? And you know what they say about a camel (the creature who goes long distances without water or peeing), "If you let a camel get his nose under the flap of your tent, the next thing you know he'll be sleeping with you.' As a former consultation-liaison psychiatrist I was an old hand at working my way into a reluctant hospital. And I have the Canadian Red Cross/ Red Crescent Training Director, Lynda, on my side.

When I was in my small tent, I had an excuse not to do all my exercises, but now I'm in a big one and in full swing (Kurtz and Joseph Conrad would be proud of me). But the hardest thing I have to do each morning is put on my knee-length support hose. I have to wear them all the time when I sit a lot to prevent blood clots in my legs (like on an air plane). Even in the heat, they are essential because of that pulmonary embolus I had 3 years ago. But putting them on I really work up a sweat. I'm no spring chicken any longer. The only thing that gives me courage, and a little smile, is imagining Patti trying to put on her panty hose in my humid tent. Though a great hiker, she wouldn't be caught dead in a tent any longer. We all have our strengths and preferences.

Remember, though, I was lying there, cozy in my sleeping sack, but it was still dark out. My mind raced through myriad thoughts and anticipations, magically thinking I was 20 again--still invincible, high and mighty. Suddenly, I felt something crawl across my feet. I screamed and scrambled to my feet, kicking off my sack, grabbing my flashlight as I jumped, quickly shaking the bag out on the tent floor. And there it was, lying on the floor for me to see—nothing. Nothing at all. Just my unconscious laid bare. So much for invincibility. When it got light enough, I also shook out my shoes, before putting them on, like my anthropologist friend, Bill Davenport, an African specialist at Yale, advised me to do, to kick out any spiders or scorpions from their new home. I have to admit there weren't even any scorpions in Haiti—that I knew of.

Being an older gentleman, 70 now, I discovered that I was living much closer to my body than in the prime of my youth, when I was in Haiti before. My pill organizer is proof, both in quantity, and memory need. But there was something about being in Haiti, and living in a tent, that was forcing me to be even closer yet to my bodily functions, or lack thereof. What I hadn't anticipated, though, was how much closer I would be to my unconscious, with reminiscences of my youth and intimations of my approaching mortality.

Being 70 changes ones planning horizon, among other things. I should mention two other personal notes, so that you don't feel I simply have morbid preoccupations. My mother died of a pulmonary embolus (PE) after a minor heart attack when she was fifty-nine, giving genetic significance to my own recent PE. And my father died at 71 from acute leukemia that hit him out of the blue, just after crossing into the 70th decade like me now. These late life anniversary experiences set the stage for one's own sensitivities and premonitions when crossing similar date boundaries. The emotional work of mourning and facing one's own mortality, while traversing and moving beyond the death dates of one's own parents creates background music for anyone doing disaster work later in life. It also allows us to bring something special to the table with patients suffering profound losses and fears of mortality.

Despite all these early morning reveries, I really loved my spacious Shelter Box Tent, given by the International Rotary Club. My father was a Rotarian all his business life, and now I was finding out about some of their great work around the world, and in Haiti. I saw Shelter Box tents all over, and it made me proud. It was also because of my father I bought the flashlight I was using. He taught me to camp, and had this funky old flashlight cranked by hand. No batteries. It never ran out of juice—if you had elbow grease. This new one was cool, much more compact and high tech, not requiring constant cranking because of an internal rechargeable battery. I looked forward to giving it to my son, Christopher. Anyway, inside my tent I was spread out all over, my insufficiently washed, sweaty clothes hanging everywhere to dry. I loved it. And couldn't wait to be done with it. I was not counting the days to departure—yet.

All this is the flip side of what IMC saw in me during the day--the old 'hair shirt', Creole-speaking 'psycho' anthropologist dubbed 'Ti Fou' who went out and worked in the mid-day sun. Now IMC had me doing it again. My English roots I guess. So these were my confessions. But it made my next confession even more bizarre. The rumor that I would have to give up my tenting life 'in a few days' when we moved to a more civilized Residence, with a bedroom for me, put me in mourning. My only defense was something like what happened in The Bridge on the River Kwai. You know, those WWII army guys taken prisoner, forced to work on an enemy bridge, and then weren't able to make themselves blow it up when they were supposed to as their own troops came to reconquer the ground. Misguided pride in ownership I guess; or, as I mentioned earlier, in Woman of the Dunes--that leprous woman living in the sand pits with the other lepers, who, when cured and about to be freed, couldn't bare to crawl out and leave what she hated and loved. How could I leave my tent?

Miragoane is an hour by van, over a road full of ruts, the roadside lined at times by beautiful banana trees and sugar cane fields, with a stunning backdrop of verdant crinkled denuded mountains, a deceptive sprinkling of green scrub growth in a few places. Like the rest of Haiti the scarce trees have been mostly chopped down to make cooking charcoal reflecting folkways and overpopulation pressure.

Looking ahead from the van I saw throngs of people, collapsed buildings, packs of dogs, and goats attacking burning refuse searching for bones and fruit peels. Gaily panted trucks and cycles hurtled toward us, with tent cities rushing by on either side.

The road was again periodically scared by those zigzag crevasses, and deep cleavage drop offs, stunning reminders of earthquake forces scarring our poor Haiti, each mercifully requiring us to stop our headlong rush from time to time—the new Haitian Zeus equivalent of speed bumps. Speaking of these, because tent camps were all along these mostly dirt roads, traffic kicked up horrendous clouds of dust. People in these roadside locations counterattacked by creating big, makeshift mud speed bumps stretching across the road. Only the animated conversation with the Hopkins Doc and Nurse made me forget the life, and death, and chaos teaming around us.

Pulling up to Miragoane Clinic, I hauled my red backpack up the steps to a building bursting at the seams with Haitians camping out patiently in anticipation. I had been told there would be plenty of chairs for my mental health clinic in Miragoane, only to find chairs scarce. They tried to put me out among the people in plain air, but I scouted around and found a cramped back room, moved soiled instruments and half-empty bottles of medicine, finally scrounging up three bent rusted chairs and a bench.

I was in heaven. Except no ceiling. The wall went up only 9 feet to a high airspace transmitting the hubbub from the next room. And there was no door. Good for air conditioning not for privacy.

After getting oriented with my Haitian doctor, Dr. George, the chief of the clinic, we saw our first patient, referred by one of the Hopkins gals. Suffering from earthquake losses, and quake shock, her anxiety repeatedly deepened by aftershocks, we prescribed some Diazepam to take the intense edge off, then demonstrated anxiety reducing exercises (those seminar progressive relaxation and breathing techniques). We added the 3-breath technique, where you take a deep breath, hold, then take it deeper, hold, and then as full as you can, and hold. You do this three times, concentrating on your breathing. Doing this is incompatible with remaining anxious. We also prescribed homework, asking her to discuss her quake experience with family and friends, their listening and sharing their own stories helping to detoxify her memories, reconnecting her to people, and 'reintegrating her narrative'. Research has shown that getting progressively more comfortable telling a traumatic experience heals and reduces the risk of chronic PTSD (Post Traumatic Stress Disorder).

In the midst of this, a toothless wizened old man, drunk as a coot, came rolling into the room giving us all high 5's. He was to be our next patient, but inebriated and high, he had jumped the gun--his poor impulse control written large in the breeze, along with his rancid alcoholic breath. When we saw him a moment latter, he was infectiously delightful--all a sad deception. As he raved on, a tear dropped from one eye during a fleeting mention of losing a family member in the earthquake, covered immediately by gay word torrents. He told us he had been drunk most of the time for 8 years, and that it was his sister's fault. She had been a raging alcoholic before him, until she saw a Voodoo priest who, for a sizeable fee, removed the devil drink from her and put it in him. I helped Dr. George explore all the awful sequelae of such chronic drinking (black outs, the DT's [Delirium Tremens, famed for its kaleidoscopic 'pink elephant' hallucinations] and Wernicke-Korsakoff syndrome [with its loss of memory, confabulation (self-deluding false story telling), etc]. Miraculously, he had been spared. I had him do a careful mental status to see if our fellow, besides acute intoxication, had the hint of other brain damage. He seemed pretty clean, to our surprise. We were naively hopeful. But hope is important.

Then I asked if Dr. George had seen his brief tearing. He recalled it, so I asked him to explore what lay behind this fleeting hint. A lot of underlying isolation and sadness emerged, which the patient usually camouflaged by his hail fellow well met veneer. Picking up on his sister's exorcism, I said to him I knew about Voodoo, and had nearly taken the *Ason* (the priesthood) myself. With a knowing smile, I said we would be willing to receive his drinking devil if he wished to give it up to us. But, we said, we could not give him a proper examination for diagnosis and a path toward cure unless he were sober. Looking him in the eye, the Haitian doctor asked him if he could try being sober for the upcoming week, to get his body and mind ready for our next visit? He accepted the challenge after we sympathized with his underlying loneliness, for which, we felt, he was taking the wrong, self-prescribed medication (alcohol). He agreed to come back. We shall see. Chronic alcoholics are tough, especially in this environment, and yet this guy had pluck, and the obvious available mental hook revealed by his tear.

Then a lovely healthy-looking but somber young woman walked in, complaining of insomnia, palpitations, visions and voices, but of a very particular kind. The voices and faces were of fellow medical students and nurses who had been trapped together with her in the medical school basement as their building collapsed on top of them, there in Port-au-Prince. Trapped in pitch-blackness, pinned under rubble, she could hear the voices, the screams and cries, of those injured and dying around her. Over four grueling days she heard these voices of her friends, kept picturing their faces to keep herself going, only to hear those voices becoming fainter and weaker, and finally dying out, leaving her alone with only one friend's voice, somewhere way up above her in the rubble darkness. This faithful friend knew she was down there somewhere below her, and told the rescuers, when she heard them above her on the third day, telling them she was alive down there somewhere below her, guiding them in both their directions. Then her voice, too, became weaker, and died out, leaving our patient utterly alone.

At first her voice was too weak to call out on her own behalf, though she could hear them calling her name. Finally she found the strength and called just once, loud enough to be heard. But the rescuers found that the pieces of concrete over her were too big to be moved. They told her they wouldn't give up, but they feared they wouldn't be able to do anything soon enough, telling her she should hold on as long as she could, and they would do their best. She lost hope though, hearing rescuer voices growing faint above her, as she hung in darkness, no sense of day or night. Her throat was parched, and her loneliness deafening, but she didn't give up. She felt she had to survive. She was the last of all her fiends. Then, finally someone got to her feet. We found out, at that point in her story, that she had been suspended upside down the whole time. As she talked with us, encouraged to open up her darkest hours, her voice grew stronger, calmer, and more certain. Her awful narrative was becoming coherent as we held her words in our open hearts. I finally blurted out I was so proud to have someone like her becoming a member of our profession.

She broke out in a radiant smile, and told us she was hoping to go back to medical school when classes started. She would be finding out the next day when that would be. But that was also the problem. She already knew two thirds of her class of 45 had died, and confessed she was petrified about going back. She was having palpitations and hyperventilation, with near panic attacks whenever she thought about getting near the collapsed medical school building again. She dreaded finding out if even more had died, and wondered about the teachers. Dr. George and I gave her some diazepam to take the edge off her anxiety and help with her insomnia. We also showed her three desensitization and behavioral techniques which would give her ways to systematically move toward mastering her feelings of fears, reducing her thoughts of impending disaster, and managing her phobic avoidance of her school and the future.

As we went over these techniques we found out she had been a student leader, and suggested she might be a good teacher and leader for student groups with whom she could share her experience and techniques for their shared anxieties, helping classmates to resolve their symptoms. Facing her own understandable feelings and reactions, by harnessing her robust strength, might allow her to refind her support community and show them how to work together, to resolve their fears and accomplish mass mourning. By the end of the session we had a sense she would be able to make it, and help shed light on the darkness they all faced. We asked her to come back with a journal of her homework accomplishments to an appointment at our next clinic, a week hence--if she didn't remain in Port-au-Prince. We told her we all felt she, especially, would be able to make it. We clarified issues around survivor guilt, emphasizing that she was living for herself, and that her self-exploration and healing would allow her to be a fine compassionate doctor sometime quite soon.

As I wrote this, tears were streaming down my face. My writing was truly therapy for me. Doing this kind of work is often like doing surgery without benefit of anesthesia, as I've said, and yet it is painstakingly important. But I didn't have time to decompress until I write. I admired the strength of many of the people I saw and yet didn't want to be too idealistic or naive. Or too optimistic. We did what we could, and hoped for the best, and tried to support the front line Haitian doctors taking care of incredible numbers under great pressure. But they had to take care of themselves, too, just like me, as they cared for their traumatized countrymen.

The next young man was a follow-up. For the second time in his life, he had lost close friends, though unscathed himself. But he had ended up not being able to hear very well. Voices seemed far and faint, if he could hear them at all. He was having hysterical negative auditory hallucinations, basically losing hearing ability because his school had collapsed and, on a deep automatic level, he needed to keep from hearing all the horrible things he had heard. So he virtually gave up hearing all together. He had heard the voices of fellow students below him, the injured screaming in agony—voices he kept hearing in his dreams and waking mind's ear. He was plagued by nightmares, which constantly awakened him. He desperately needed to get rid of all this so he could stand his own mind and not be held hostage, or driven crazy. He wanted to be free to pursue his life. But the cost of his hysterical protection was severe, not being able to hear other things in order not to hear these anguished cries.

In his follow-up visits with Dr. George, this being the third, he had already begun to hear better through his continuing support anti-anxiety medication. As memories and feelings came back, he was flooded with painful but laudable grief. In coaching Dr. George, I tried my hand at interpreting in my 'Franc-Creole'. I put into words my sense of his torturous mental journey, try to instill hope the road ahead was possible with the end point 'in-sight'. It was particularly gratifying to work with Dr. George. He was the doctor who had come up to me during my first seminar sharing his own losses. I appreciated his sensitivity.

What were our thoughts? Dr. George and I discussed the fact that this bright timid, inhibited patient had been able to get through similar symptoms successfully around his first, pre-earthquake loss. I pointed out that this gave us hope, and a predictive model, for recovery, (a good prognosis), though we emphasized he was carrying and bearing a lot more this time. His more child-like early self-protections (defenses) were giving way to healthier, more mature ones. But there was a price, requiring courage. Painful but healthy waves of grief and guilt began to wash over him as he dealt with his losses. We urged him not to be ashamed of tears, but to journal his efforts, and have the courage to share his sadness. We asked him to come back in a week to help us appreciate all the hard good work he had done. We also discussed the burdens and pitfalls of survivor guilt. The patient left confirmed in his continuing work, and his painful progress. I was proud of Dr. George's work with him, and shared my admiration with him.

Our last patient was a cute little girl, with severe developmental delay from birth, who had seizures and had lost her medication when her house caved in. Her doctor had been injured and was unavailable. So she needed to get her seizure meds from us. We breathed a sigh of relief at such a routine request, which we filled with pleasure. We determined in the process that her medications were not controlling her seizures very well so arranged to adjust them and have her come back to make sure we got them right.

My gifted interpreter, Tessier, a schoolteacher out of work because his school had been damaged, turned out to know a number of these patients and their families. After the Clinic he confessed he felt dizzy and drained, feeling sick to his stomach. We both commiserated about all we had heard, agreeing it was a lot to swallow, especially with open hearts and ears. We both needed some R & R. And yet he felt he was privileged, and learning a lot. And he had noticed something that had pleased me a lot. The head nurse, who had been at my opening seminar, made it a point to come and sit in on our work, paying dividends for her and for the clinic, benefiting future patients. We all looked forward to meeting again next week. I stuck my head in and gave the Hopkins doctor feedback on the patient she had referred to us.

My shirt was drenched a good part of the time. My best self-care, though, was the fact that I had sprung for a blow-up camp pillow, which at first I was embarrassed to take out, until my seat couldn't take the rock hard chairs any more. So I would blow it up, soon making it a ritual. Tessier and the doctors, and the nearby patients, especially the kids, loved to watch. And boy was it comfortable during those long grueling sessions, during which I had to have the quickly spoken Creole translated to me, and then my words fed back in Creole to the doctor. At the end of each case I tried speaking some in Creole to model how to give interpretations, at times drawing quizzical looks, at other times confirmation of my words. Often, though, Tessier had to re-translate my 'Creole' into Creole. Chagrined, I was happy to be rescued. We kidded each other that he would soon be getting his own psychiatric diploma. I bought a BIG Haitian cola from a vendor on the way out which I guzzled thirstily on the long ride home.

When I got to the Notre Dame grounds, before going into the hospital, I found the lab guy who could do my anti-coagulation test. I got my blood drawn under rather sketchy but clean circumstances. Then they informed me they couldn't do the test (an INR) I wanted, but could do a bleeding time, to see if I was at least in the ballpark of proper anticoagulation. Given my altered diet and my significantly different fluid balance situation, I was wondering how it would turn out. As I mentioned, in the Clinics I could go nine hours without peeing, and when I did, it was quite dark from concentration. Amazingly, all this was happening despite consuming two huge bottles of water during those long hours. I guess it was almost all perspired away in the sweat lodge.

I cruised by to see my helpful Canadian, Lynda, who said, "Sorry, come back tomorrow, I haven't had a chance to bring it up with my residents." I suspected she had forgotten.

"Thanks, I'd appreciate you seeing if they might want a seminar or two."

When I got back to the Residence at 5 o'clock, my belated long-cold lunch was waiting, nestled under the fly protectors, at least what was left of it, all made by Crystal the cook, including some still good red beans and rice, that vegetable stew, and a wonderful Black Bean soup. My stomach had shrunk, but not my appetite. I was feeling healthy and great. Interestingly, all my joint problems, despite sitting for long periods, had gone away. Maybe getting over being rusty was a good thing. I was doing my exercises faithfully. My other thought was, it's my daily sweat lodge treatments.

Wednesday, March 17: The Boat Clinic and the Mad Woman of Platon

I was really looking forward to the Boat Clinic the next day, and eager for a good night's sleep. But at 4:30 a wild cat screamed a mating call in my ear, after dropping into our yard to eat our garbage. Fired up by a shot of adrenaline, which stirred up my fight/flight endocrine response (from my lecture last Saturday), I screamed bloody murder back. We exchanged two spirited mating calls and the cat finally skedaddled. It was deliciously cool by this time in the early morning. I luxuriated and began drifting off. Then something big fell on my tent from the overhanging mango tree. Just a mango I thought, until it began crawling along the tent ceiling. Was it the cat? Or something worse? Or just my unconscious again. I said to hell with it, and turned over. The king of denial went back to sleep.

All I knew about going to Platon, a small remote village, was that we would be taking a boat there, along with two Haitian doctors and two nurses (and no medical volunteers).

I was up before the little guy came around to sweep the fallen mango, avocado and coconut leaves away, something he did every morning. I could hear his 'swosh, scrape swosh', as I did my exercises. I hoped that whatever was on my tent didn't bite him, and that he would get rid of it. After doing my morning ablutions, I walked out to check out the action.

The transportation guys were jabbering in Creole and revving up four big 6-seater Nissan Patrol cars and 2 large Nissan 8 passenger vans, getting ready to cart us around to the various ambulatory clinics. I had already taken part in this morning pageant, leaving early each day as I did, but had to ask Samedi, the dispatcher, where the Boat Clinic people were. Now Samedi is cool, and from early on we've had a little thing going. You see, Samedi, which is French for Saturday, is also short for Baron Samedi, one of the most powerful and feared Voodoo gods. So I called him Baron at times, and he and the drivers cracked up. But it was no joking matter. The Baron is god of death and the cemetery, and also head of the Bizango Society, which enforces community values, often with frightening summary judgment. When I was here in the 60's, Papa Doc Duvalier, with his feared machete and machine-gun toting Ton Ton Macoutes, would actually dress up publicly like Baron Samedi, all in black, with cane and wire-rimmed glasses, assuming a cold, poker-faced reptilian stare. Over a Barbancourt rum one evening, Jean Blephus Richardot let me know that Duvalier actually held Voodoo ceremonies in the Palace and had his own in-house *Houngan*, or Voodoo Priest. None of this was lost on his Haitian subjects, instilling fear and respect in them.

Our dispatcher, by reassuring contrast, was a short, wiry good-natured guy, easy to work with. But he liked my implying he had powerful administrative back up. Anyway, Samedi told me the boat people got started a little later than the rest, because they always waited for the two Haitian doctors who often arrived on "Haitian time, not American or IMC time". Then Samedi pointed out two boat nurses who were already there waiting. They smiled at me and I did a double take. They were the same two nurses from my Saturday seminar who finally admitted they had no home or tent, no shelter at all. This was already shaping up to be quite a day. Little did I know.

When the doctors arrived, I recognized them, too, and we were off to a good start. Piling into the van, we bounced down to water's edge, passing the ruin taken over by goats, going just one street past the turnoff for the Royal Hotel.

When we arrived at the beach, already teeming with fisherman and guys mending their nets. I took out my camera and walked past old, brightly painted dugouts and the bright yellow fiberglass runabouts. Picking my way through the refuse and rocks, I realized I had seen this place from the Royal 'beach'. I began swinging my camera around to take a picture of the beautiful azure bay and the distant mysterious Il de La Gonave, destined to be in my sequel to **Body Sharing**.

As I looked out over the water memories came flooding back. On a similar, stunningly beautiful, cloudless summer day I had taken a voyage to that island 50 years ago with Haitian peasant fishermen in their rickety sailboat. We arrived around noontime at their own personal off-shore island, made entirely of conch shells they, and generations before them, had caught and laid down. We had already said goodbye to their on-shore wives in the cool of early morn.

Now they were introducing me to their second wives, their 'island' wives. Haiti is a polygamous society, IF a man can afford it. And these guys, with their thriving conch, or *lambi*, fishing, were in good enough shape to pull it off. I went skin-diving to my heart's content, and came kicking in for a surprisingly good conch-stew dinner.

On the next day, I went out with them to the prime *lambi* hunting grounds to see how they caught them. I wondered what all the long sticks were for, and found out they lashed them together, to a length of 35-40 feet, placing a bamboo chock on the end, something like we old codgers use to pick up things on the floor or reach up to unscrew ceiling light bulbs. I watched them use glass bottomed buckets to pick out big lambis on the distant sandy sea floor, and then unerringly drop the pole down over them to chock them, keeping the boat steady in the process. I was foolhardy enough to bet them I could dive down and get one, not realizing how deceptively deep it was, given the crystal clear water.

On my first dive I barely got down half way, and came hurtling back up for air. But I was youthful, foolish, and undaunted, and after 'breathing them up' 'til my head spun and my nose tingled, I dove down down down down and grabbed a prime lambi, forgetting I had to come all the way back up. I felt a tearing pain in my abdomen and shot back up, nearly passing out, but holding on to my prize. They were impressed, but also cracking up with my breathlessness. I had a stomach ache then, and later acid reflex, finally back home having a barium swallow and finding out I had torn my diaphragm a little. The pressure collapsed my lungs and forced the neck of my stomach through my diaphragm. So much for youthful prowess and vanity. Even then I turned out to be vulnerable. But at my age now I think I'll stick to having my lambi in the buffet line. Easier to reach.

While we waited for the boat to be ready, I told this story to the doctors and nurses, skipping the stupid parts, and some fishermen overheard me and laughed. Why do fishermen always laugh at me? They probably didn't believe me, and for good reason. Let me tell the finale to my ancient story. As we finished lambi choking, filling the bottom of the boat, the fishermen and I set sail for the distant coastline and home. I heard them muttering something and saw them pointing toward the horizon. An ugly looking front was just coming over the tops of the mountain range off the southern peninsula just behind Leogane. I was enough of sailor to see big trouble coming.

We had gotten only about half way back to the mainland when high winds and sheets of rain struck us. Jagged streaks of lightening crackled all around. We had to furl sail and tack widely. The boat was already low in the water with all the lambi, but I got the sense they'd throw me overboard before those prized conchs. The boat and its rickety rigging began singing, and as the wind pressure mounted the pitch increased; finally a few lines began shredding and one broke with a loud pop. We finally had to drop sail, turn tail, and just run with the wind, threatened by huge after-following seas. I hunkered down as it got dark, not just from the storm but the cool of approaching nightfall. I was trembling. I guess I was more scared than I admitted. I thought of all the Haitian 'boat people' and African slaves lost to the watery deep. Suddenly I heard one of the fishermen break out in song. I peeped out to see why. The moon was breaking through the last of the passing clouds and the lights of Leogane and Ca Ira were sparkling in the distance. The ugly squall had passed. We had lost a lot of ground, but not our lives, and I had had an experience I would never forget.

So here I was, looking out at the mysterious Island of la Gonave, partially shrouded in morning fog, a few clouds shrouding the far end. The Clinic boat crew were having trouble getting the engine started, and I had trouble getting into the boat. Why hadn't I listened to Patti when I was packing? Just for once. She had handed my water shoes in case I went swimming, and, 'know it all' that I am, I willfully left them under a pile of dirty clothes back in Paris. *Too much weight and something I'd never need*, I thought. Well, as we walked down to the boat, which was several feet out in the water, I noticed all the staff had on water shoes, of one sort or another. Tessier was in the same fix I was. We finally took off our shoes, rolled up our pants, and walked out through the rocks and water, crusted with refuse, the flotsam and jetsam advertising many brands. Tessier took off his socks, but there I was, with my goddamn support hose. No way would I entertain everyone by my struggling to remove my support hose. My bizarre contortions were strictly a private matter. So I just walked on out. They dried quickly, except for the fact that, under way, we headed into swells that splashed me periodically. I was grateful, because the morning chill rapidly gave way to increasing warmth as the sun beat down on us.

Boatride to Platon, tentless nurse Eustache middle, Marie L, Kent R, Dr. Beauge R behind

The boat ride was an hour, and for security reasons, Tessier and I had to wear life preservers. I sat in the middle with the nurses. The docs sat fore and aft, unpreserved and enjoying the breeze—and missing the spray. They knew where to sit.

None of this mattered because the boat trip was amazingly beautiful. We pulled out past Royal Point, cruising across the outer reaches of Petit Goave harbor. Farther out a huge Spanish army hospital ship was moored. Behind Petit Goave the green undulating foothills, deep ravines and verdant valleys of the foothills gave way to the high mountains forming the spine of the southern peninsula. Two tankers, one bright red with a rusty water line, the other a dirty streaked white, were anchored on the other side of Petit Goave Bay. Flying fishes skittered across the water. One of the nurses talked in animated fashion to Tessier describing everything that happened to her and her family during the earthquake. Jutting out majestically on the other side of the Bay was a high mountain dropping sharply into the azure waters like the side of a fjord.

The mountainous spine of the peninsula continued out of sight behind it. Haiti means mountainous in Arawak Indian, the language of original indigenous inhabitants.

Looking at the mountains, I recalled another youthful adventure. To the horror of my peasant family and friends in Masson, I set out early one morning planning to hike over that spine. From my Leogane field site I marched straight up the Momance River, cutting off on the Orangier tributary, bushwhacking along its edge until I found a well-worn trail, finally reaching cool breathtaking heights of Morn Campon, near the crest of the spine.

The quiet path I had chosen led me toward a solitary summit, perfect for private communing. Ah, to have a moment alone in Haiti. But it happened to lead to a market place at its summit. Near the top, I took a dip in a cool stream by the side of the trail, lying back for a second on the embankment. I woke up half an hour later surrounded by curious peasants. Gulliver had been discovered. One is never alone in Haiti. Even so, I was in heaven, until a Haitian Chef de Section arrested me, suspecting me of being a white drug dealer. Someone must have told him a suspicious 'Blanc' was asleep up there. The Chef de Section was on horseback, and motioned me to follow him (and his horse). My protests got rather loud. I brandished my Letter of Identity and '*Sauf Conduite*' defiantly. He wouldn't budge an inch, and forced me to follow him behind his horse, like some medieval thief in tow. It turned out he couldn't read, especially flowery French from the Haitian government. So there I was, being paraded through the market place, to the laugher and delight of the Haitian peasants, who formed a moving throng around and behind me, following along and pointing. Some even mimicked my outcries and uppity behavior. I was the entertainment for the day. I suspect things haven't changed much some 50 years later.

The guy led me all the way down the far side of the mountain right into my intended destination, Jacmel. Saved me from more bush-whacking and getting lost, I guess. When we got to the Jacmel police station, they finally found a top officer who was able to read the 'Sauf Conduite' (safe passage), apologized for the mistake, and let me go. I thanked him for my guided tour, and asked if there was somewhere I could stay. He pointed me toward the Pension Alexandre, and off I went. Sitting parched and exhausted on their terrace overlooking the lovely Jacmel harbor, I spied birds in a cage, and found out they were dinner pigeons, to be selected like fish or lobsters in a tank. I had never eaten pigeon, felt a little squeamish (and guilty) for a moment, but then chose a big plump one, enjoying the first roasted squab of my life--an exquisite experience. I was content, but underneath I felt a little cruel and self-indulgent.

As our Platon Clinic boat bounced along, we skirted around the rocky promontory, and soon saw some huts surrounded by palms and banana trees. "Bananier, one of our Clinic sites," yelled a nurse.

After passing another site, Goumbe, we plowed to a stop at Platon, a large isolated fishing village. We had passed a number of boats on the way, some picking up wicker traps, marked by plastic Coke, Orangier, and water bottles. A stray Perrier bobbed by. I realized most of the dugouts came from Platon, though at this early hour many were still pulled ashore, along with a few primitive sailing craft.

The purr of our motor attracted quite a crowd to the beach, children running down to greet us, the adults hanging back to keep their place in line. I discovered the Platon Clinic was open air, shaded by huge Tamarind trees, and a scattering of coconut, banana, and mangoes, many heavy with fruit. Pigs, chickens and goats had the run of the place. There were perhaps 50 patients standing or sitting, many with children, some nursing babies.

I was armed with chairs this time, asking where I should open my office. One of my nurses, Marie, looked around carefully, choosing what she felt was a prime shaded spot behind the sole village building, under a truly majestic Tamarind. A second tree shaded a boat carpenter cutting and shaving planks. Ringing him to enjoy some of his shade were a half dozen people, chatting and loitering about. When my table and chairs went up, the group swelled in anticipation.

"So how many patients do you have for us today, Marie?" I asked.

"Nobody yet, well one, the lady Eustache is bring there," she said.

On impulse I asked, "Are there any really crazy people around?"

"Yes, three." "Round them up, if you can. That's what we're here for."

She disappeared for a minute, and then came saying, "Somebody will show up shortly."

Meanwhile, Eustache, my tentless nurse, brought over a rather sad looking lady with twins (see the front cover of this book). She had lost her husband, and three of her older children had died of illnesses, unrelated to the earthquake. The tremor had taken her house, and her hypertension medication, leaving her without family or shelter, and mounting blood pressure.

 As we were just adding Atenolol to her other blood pressure medication (also good for her tension headaches and hyperventilation), we heard a commotion coming down the trail to our left.

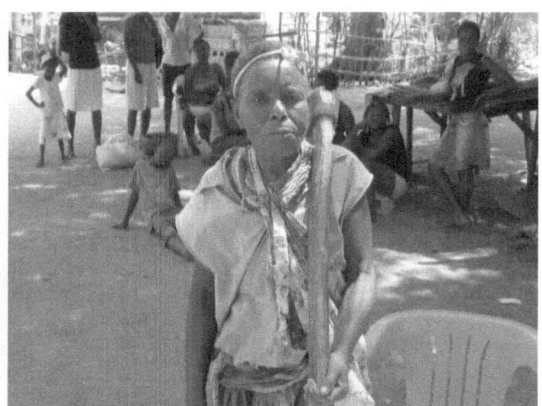

Carrying her hoe ax for security, our second patient marched up

This wizened old lady came roaring out of the banana trees into our office clearing, a relative trying to keep up with her. She was dramatically shouldering a hoe-ax and looked like the Voodoo god, Cousine Azaca, with her tattered dress and scraggly rucksack slung over her shoulder. She marched straight up to our table, with a retinue of onlookers, not really in a threatening way, just with dramatic determination.

To our shock, and the delight of the crowd, she threw up her dress, pulling it over her head, showing us her emaciated body, sagging breasts, and. oh yes, her privates, emphasizing in Creole how thin and hungry she was. "I have no food, no shelter, and my relatives have all abandoned me, and look, the village all make fun of me when I'm crazy like this. Please, could you build me a house and give me some food!"

The onlookers roared at her outlandish request. We settled her down, saying we would see her right away next, but asking her to wait her turn. She sniffed, huffed, and wandered quietly back into the crowd. I watched her movements, her shifting mood and antics out of the corner of my eye while finishing with our current patient.

When we finally saw her, the crowd wondered if she had an appointment, then pressed in around our table to listen. I felt badly about the invasion of privacy and tried at first to clear them out, barely quieting all the laughing gawkers. Then I realized she was doing theatre and so were we, and from what I saw, we would need to involve family and community as part of helping her, with the hope they might re-accept her—if she calmed down. After hearing her sad story and long downhill course, I had a hunch. I asked Dr. Beauge to take her blood pressure, which came in at a staggeringly high 210/110. This made it clear she had, at a minimum, a fluctuating hypertensive encephalopathy, and a probable *Etat*

Lacunar, her brain pock-marked by myriad micro-stokes, which had eaten slowly away at her mental capacity.

So we had our diagnosis, (unless she also had pellagra or something else from malnutrition). But I always followed Occam's razor, to begin with, looking for one unifying diagnosis, rather than a complicated, multi-diagnosis approach. We explained our thinking to her and arranged to give her antihypertensive mediation. We medicated gently taking care not to cause her a low blood pressure 'watershed' stroke. She probably had some vessel narrowing and low pressure would close them. Our simplified explanation helped her settle down, and served to inform the family and eavesdropping community about what was going on. We were also doing a little public health work, informing the community about the risks of high blood pressure as we went along. But mainly we wanted them to understand her plight and treat her with more respect. We asked her to come back next week with her helpful neighbor.

As we were doing the next case, she came cruising back by, holding out her hand with a few beans in it, asking if we could give her a few more so she could have enough to make dinner. Everyone laughed and we found ourselves smiling too. She looked sternly at me and said, "I won't leave until you agree to build me a house!" I was kind but firm, saying we would look forward to seeing her next week, but the medication was all the help we could give her right now.

The nurses were watching and learning, and the docs found this and other cases very interesting. The boat ride back was great. I noticed some conchs, crabs with big claws, and a few choice lobsters in the bottom of the boat, prize purchases by staff from the local fishermen. They looked so good I had the thought, *Lobster tonight at the Royal!*"

The thought sparked another memory. My first Yale summer in Haiti I was still torn between anthropology and medicine, but found myself drawn into premedical practice. At first I went down to a community compound called Louis Tore, treating children for Trachoma and other eye infections with eye drops. Hearing about it, people wanting help began coming by Ternvil's where I lived. One of the fishermen I had sailed with out to Il de la Gonave came by one morning, asking me if I was willing to come down and see his ailing mother. Her bloody cough and emaciation reeked of tuberculosis, so I made a mental note and picked up TB drugs in Leogane the next day. When I dropped by to get her started, his son was also feverish and hallucinating, and had been for several bouts that week. Because the fever and chills came and went in a cyclical pattern, it was easy to spot his malaria, something I knew only too well. I came back later that day with some of my own Chloroquine, having read the package insert for the child treatment dosage. His mother improved slowly, and his son was better overnight. Two days later, Ternvil called in to me as I was typing my field notes. I looked out the door, and standing there was my fisherman friend, holding up two huge spiny lobsters brought over just for me. Deeply touched, I thanked him warmly and took my lobsters, careful to hold them by their antennae. I showed Joselia how we boiled them in Maine during my teen summers in Tenants Harbor. She made me a feast for a king. I had already lost 15 lbs by then, and was ready for a Maine meal.

Not all my medical endeavors that summer ended so nicely. Ternvil's next-door neighbors brought in their badly burned little daughter. Hot cooking oil had spilled accidentally on her inner thigh and calf causing deep second and third degree burns resulting in huge skin blisters over a third of her leg. She was already feverish and delirious. I had worked as an orderly at Barnes Hospital in St. Louis and helped hold patients being treated on the Burn Unit. I remembered they used a sulfa-based medication, Silvadene, on bad burns. Since I was going into Port-au-Prince the next day, I picked up some Silvadene. I worked with them for two weeks using Silvadene, and adding a systemic anti-biotic by mouth, and leg warping, keeping the burn moist and covered. By the end of the second week the wound was doing well, with the new skin buds coming in nicely. It can look like there's no skin there at all at first, and no hope of regrowth. But deep down in the hair follicles are primordial skin cells which thankfully provide the roots of skin bud regeneration.

Since I was often preoccupied with other things, if I forgot they always came by to remind me to come see her. Strangely, one day, no one came by. Then I noticed quite a crowd at the house. I strolled over. Standing in the middle was the local herb doctor (a doctor *feuille*, or *bokor*), who had ceremoniously unwrapped the leg and poured a drying powder, a special dust, on it. I had been eased out without a word, and the *bokor*, with perfect timing, had taken over my case at the time of success. I was miffed but felt the baby girl would be all right. I also knew the dry mud powder he was using probably came from a sulfur spring I had heard about up on Morn Campon and would work okay. Silvadene is a sulfur medication as I mentioned. So my first Haitian consultation-liaison case bit the dust (successfully).

When I arrived at Notre Dame Hospital to do my consultation work that afternoon, I was hoping to have better luck this second try at consultation-liaison work in Haiti. I began making informal liaison rounds, met with Dr. Lynda again, asking about any possible seminar interest. She still hadn't asked the residents. Frustrating, but familiar to liaison psychiatrists. But I knew I had to keep waving the psychiatric flag. Something would happen someday. Just after we finished talking, I got an emergency call through Stephanie from Croix Rouge about seeing a case of a mute traumatized girl, and made rounds all over the hospital searching for this in-patient, only to find out she was an outpatient and the problem had been solved already. But through this process I got a little legitimate liaison visibility going, meeting a delightful energetic Norwegian Red Cross coordinator, Brita. Waiting for my driver, I had a few minutes and went over to check my own bleeding time results. They didn't have the results yet. So I had to continue 'flying blind', not knowing whether to increase or decrease my

Coumadin. All I knew was my teeth didn't bleed when I brushed them, something I had learned was a sure sign I was overmedicated. I began worrying they had lost my blood or didn't know what they were doing. But I was sure I was fine.

A Dr. Jeannot Francois, the Director du Department Sanitaire des Nippes region called Joanne to reach me. I called him back, and he said he would like psychiatric help in the Miragoane Hospital, and could I do it. I said I was pleased with the prospect, but was just a volunteer for the next 2 weeks, and would pass his request on to Dr. Peter Hughes. How many fronts could one guy open up, anyway? The call made me a little anxious, but I was comfortable saying no and passing the buck. This represented an opening, but Peter was spread even thinner than I was. Fame can lead to flameout.

That evening, I was sitting across from Stephanie doing my diary, writing about the Haitian medical student buried upside down with her classmates. Stephanie looked over and said, "What's wrong? You're crying"

I read her an excerpt. Her eyes began to glisten. "I had no idea you got so involved. You really care." Then she got very quiet for a moment. Finally she said, "It's my fault the internet's been so slow. I've been making long Skype calls to my abducted friend. They just released him."

"Thank God. Must have been awful."

"You have no idea. I never told you. I was abducted by the same people in Darfur. When they let me go, I warned him they'd go after him next. He spoke out too much about the rebels. The abductions are politically motivated from the top, but the guys who carry them out are in it for the money. They're ruthless. I had one moment when I really thought they were going to kill me. It could have happened."

"Yet you've come back to do this work again."

"What you said about helping that girl be brave and go back to face her fears so she could carry on, that reminded me of how hard it was for me to come back to work. I took time off after I was released but I was in denial. When I decided to go back to work, that's when it hit me. I was incredibly anxious and stuck. Talking to someone really helped."

"But you made it. Talking helps. Your friend's lucky to have you."

"Thanks. See you tomorrow," she said, throwing me a warm smile. The ice was broken. I admired her.

As I walked back to my tent, it began to rain. It was nice crawling onto my cozy air mattress, lying back to listen to the rain pattering down on the tent. I felt a peaceful calm come over me, and drifted off to sleep. Several hours later I was awakened by a loud drumming. The rain was pelting down in torrents now. Suddenly I had the awful thought that the rainy season was upon us. I remembered the pools of water in my tent when I first arrived. Flicking on my trusty flashlight, I beamed around the tent floor, relieved to find no significant water. And no tarantulas. I drifted off again, and I had a brief dream. I was floating down a river on my mattress. The rain racket woke me again, and suddenly I felt sad and guilty. Here I was enjoying myself and worrying about a few puddles, and all these Haitians were out there getting soaked, many in flimsy makeshift tents, with the water pouring through and drenching them.

And what about the tentless nurses from the Boat Clinic? What was happening to them? Oh, and the mad woman of Platon, the one beseeching me to build her a house and give her food. What about her? As I thought about it, I suddenly realized I had missed the boat entirely, since the primary mantra of IMC was to take care of food, shelter, water and security first, the basic things that help most people through the crisis. We had laughed at her pleas without attending to them. And what about dietary deficiency diseases, possible causes of her dementia? Giving her multivitamins would be easy and cost nothing. Yes, we had missed the boat, entirely.

So I made myself some dreamy promises--to dispel these guilty nightmarish worries. I determined to give her Norbert's tent, going unused in the corner of my big Shelter Box palace. And to get Crystal, our cook, to buy her some rice, beans and cooking oil. I could certainly scrounge up some vitamins. Yes, there were things I could do. "Yes we can! popped into my head. Except we weren't supposed to do direct giving like this, partly because of the problem of stirring up envy and competition. Luckily things were much better between Stephanie and me, and I could ask permission to give these things to her quietly, with medication written in magic marker on the tent bag. Finally, the rain seemed to quiet down a little, after drumming some sense into my head.

But it was still raining when I got up. I floated the idea of the tent past Stephanie in the office that morning. She was okay with it. I was relieved. I went outside and found the Boat Clinic nurses waiting for the doctors by their van. They thought the plan was great too, but then one of them caught me off-guard, whispering, "Don't forget about us!"

I got an email from Patti telling me about a CNN report on Haiti, saying the rainy season had begun, and describing the Petionville tent camp hillside, depicting rivers of water and massive mud slides running down between the tents, washing around, under and through them, with clothing and bedding ruined or washed away. The mud was several inches deep all over, with Haitians sloshing and sliding around in it, all covered avidly by the CNN videographer and reporter. It was the beginning of what everyone feared, an unmitigated second disaster. My worst nightmare seemed to be coming true.

Late the next day, to everyone's relief, the sun broke through again, everything dried out, and it has been fine ever since—at least for the time being. But I feared the rain might be a harbinger of what was to come, and I was worried. I've seen a few trenches around tents for drainage, but not much else. Life went on as usual, the grownups cooking and running little tent-side stands or going to jobs (the 10% who have them), while the kids flew makeshift kites and played with their dolls. I was worried the tent people were in for something awful. Then, again, what else could they do? No place to go, no government help and the NGOs overwhelmed, many slowly pulling out. That was the problem. IMC planned to stay the course, money and donors permitting, setting a goal of two years. Writing this I found myself tearing up again. I didn't want this to be Haiti farewell. I wanted it to be Haiti fare well!

Thursday, March 18, Beatrice Clinic: A Moving Experience

The tectonic plates of my unconscious were slipping again. The return of the repressed, some would say. Just chalk it up to an old fart revisiting his youthful haunts, unearthing memories while helping dig out beloved Haiti. Getting closer to death seemed to breath life back into me, as I continued to take the most important final exam of my life. And it wasn't just psychiatric, but medical, too. I've treated four times as many seizures in the past days than in the rest of my career. New cases, lost pills, dead doctors, the convenient presence of our new clinic, word of mouth, all contributed to this gold rush. Being with my Haitian doctors and patients put me closer to my mind and body because I had the privilege of being so intimately close to theirs.

Early Thursday morning, I tried Skyping a former Virginia analytic patient, now going through a momentary crisis around her newly adopted son from Russia. Unfortunately, the Internet was spotty and malfunctioning. Very frustrating. Today the connection gave my patient and me only momentary hissing word fragments and mosaic facial distortions. Frustrated in real time, we still knew we had touched base, and that was enough. But the Internet problem continued, interfering with sending diary blog attachments to people. I had to put my many words into the body of emails. 'My body to yours," as I said. Then I thought of Stephanie eating up our precious quota of Internet minutes talking to her recently released friend and I teared up, suddenly not minding at all.

On the way to the Beatrice Clinic, our driver turned left onto a narrow rutted dirt road, already crowded by vendors and women carrying heaping head baskets. Our driver dodged potholes, children, and stray goats as we wound our way up a steep hill. Honking constantly, he weaved around cave-ins, sleeping dogs, rock piles from collapsed houses, and occasional clean-up workers, who dove helter-skelter for cover. Hermetically sealed in air-conditioned splendor, with the latest Haitian rap playing, I had a sudden chilling thought, *what are they thinking of us, tearing through their chaos in a cool pristine white Nisan patrol car, a blanc in the front passenger seat. No wonder there are abductions by a 'patriotic' criminal few.* As a former anthropologist, I felt embarrassed—a conspicuous sitting duck, or Raven. But security was the name of the game these days for IMC and all the other NGOs. Hitting a pothole jolted me back to reality. I thought of my son Christopher when we did bumper cars. We broke out into the open to a breath-taking view of Petit Goave harbor below and the rugged mountains above. To our left on the edge of a big tent city stood a large khaki brown army tent, a blue and white sign reading IMC, already surrounded by hordes of neatly dressed men, women and children, their colorful shirts, dresses and scarves brightening their long wait, sad stories and urgent needs.

Saying "Bonjour" and "Ki gen ou ye (how are you")), Tessier, my translator, and I worked our way through the crowd to the back of the Clinic tent, precious folding chairs clasped tightly in our hands. Dr. Judson, our Haitian family practitioner, joined us as a harried but smiling nurse ushered in our first patient.

She was a 38-year-old woman with a constant headache and a peculiar inability to comprehend what was going on. She complained her head felt 'charged' or jammed. She had not been able to take in food, and barely took in water since the earthquake. It was several minutes before her physical ruminations abated and she mentioned her husband had died when their house caved in. When we asked her if she had ever felt this way before, she recalled a time of crisis during school, and then described what, on careful examination, sounded like pseudo seizures.

Tipped off that she could use hysterical defenses, I discussed the situation with Dr. Judson and he told her we realized how deep and intolerably painful her loss was, that she must have loved him so much, and so her mind and heart couldn't bare to take in this awful thought, shutting down taking anything in, even by mouth. We felt her mind was protecting her by being jammed, keeping the painful thoughts away and leaving her head aching to avoid such a deep heartache. As Dr. Judson said this, tears began to stream down her face, followed by wrenching sobs. I felt sadness welling up inside me, too. She had been suffering from arrested mourning. We spent a half hour with her, supporting the beginning of her painful grief, and encouraged her to have the funeral she had been avoiding. A family member of hers was encouraged to help the whole family work with her and share their mourning.

Our second patient was a 47-year-old man saying he felt awful all over and had lost his mind. He talked in a peculiarly flat halting way, sounding like a child. The quake caused the floor of his house to cave in under him. He knew his aunt was in the room below and had been crushed. For ten agonizing minutes he felt panic and fear tear through his mind and body. Then he heard her voice calling for him, wondering if he was all right. One of her legs was badly broken but she was all right, luckily having been in a doorway. But the panic stayed with him, setting off a mental chain reaction. He had been led in by another family member, who told us he had been hospitalized at age 18 when his second parent died within a year. After 6 months he was released, and had been simple-minded and helpless since then, at times not making sense, never working, and needing to be cared for. His aunt was his caretaker. He acknowledged his deep fear of losing her too, and felt helpless and passive. I realized he might be a simple schizophrenic, probably acting like a 'pseudo imbecile', who had been shocked into decompensation. I said the family needed to do a lot of 'reality rubbing' around reassuring him his aunt would be back to taking care of him, that he was not alone in being scared to death, and that in the meantime others would take care of him, especially while he got over his understandable fears. I also had doctor Judson give him a small amount Chlorpromazine, a mild anti-psychotic, to help his thought disorder, his anxiety and his sleeplessness. He felt understood and accepted, and left in somewhat better shape. We planned to see him the next week in follow-up.

As the sun beat down on the sloping tent roof, sweat dripped from my chin, my shirt soaked through. I opened my second large bottle of water. Our third patient, 22, complained of dizziness, foggy thinking, and palpitations, and a peculiar tingling in her lips, nose and fingers. We noticed she was hyperventilating as she talked of continuing earthquake and aftershock fears. She proved bright and eager to learn the self-help techniques of sack re-breathing, good for the overbreathing, tingling and dizziness, the Valsalva 'bearing down' maneuver for her palpitations, and family talk about the quakes to desensitize her fears. These interventions in her neuroendocrine fight/flight stress cycle would diminish the amplitude and inevitability of these cycles. We also worked cognitively on her maladaptive self-diagnoses of cardiac and physical illness, asking her and her family to engage in family discussions about what had happened. Through this she could regain active mastery, begin to trust her body and mind again, and unglue her stuck overactive stress response. With all this her mind and body would slowly right itself. We realized that the three people standing nearby were related, so we prescribed working together and urged her to lead the family group and teach everyone her new techniques. She felt empowered.

Our fourth patient was a peculiarly jovial, overly talkative, sleepless motor mouth, eating like a hummingbird and driving her family crazy. She was denying fears and losses and was clearly manic, riding high mentally and emotionally over her low feelings and black thoughts. So we put her on Carbamazepine, good for mania and seizures (cooling overheated brain electrical activity). We explained to her family she was riding high above their shared fears and losses, and that, as she came down they could begin to help her acknowledge and face these things, allowing her to reconnect her grieving heart with her denying mind—all part of the sad work of dealing with their shared losses. But first her protective high mood needed to cool off. Dr. George had never heard such ways of talking to patients, connecting up mood and loss and manic defenses, and tying in the use of medicine in such an integrated way. He really liked the synthetic art of interpretation, including healing words maximizing the 30% placebo effect of modern holistic medicine. He was clearly enlarging his medical mind and personal bag of techniques. It was a privilege and a joy to work with such an open eager colleague.

Our fifth patient was a 31-year-old woman with the wildest migrating shakes and quivers I had ever seen. Mesmerized at first, I finally distracted her, watching her muscle group quiver patterns more carefully. By noting they started and stopped in a self-conscious way, clearly under voluntary control, I was able to determine she was faking them. I even had her stand up so we could watch one buttock and then the other quiver like a stripper's, while other parts stopped twitching entirely. I abruptly pointed this all out to her in Creole which surprised her, asking why she was taking up our time like this and what was she really wanting from us. She got up and left in a huff, with nothing quivering any more as she high-tailed it out the tent flap. My Haitian colleague rightfully got after me for my ham-handed treatment of her, reminding me she had been through a lot and had her needs too, which we might have questioned her about more gently. I opined we were seeing such sad needful cases that I just lost it with her. I learned something from Dr. Judson that day, and I told him so. He was a kind and patient doctor, and I respected him.

We took a brief break, allowing me to walk outside to cool off a little and enjoy the spectacular view of the high mountains climbing off to the left, scarred by deep ravines and peppered with clusters of peasant Kays (thatched roofed wicker-wattle walled peasant huts with occasional collapsed stucco-walled houses).

I watched stately woman of all ages, with beautiful features and perfect posture, carrying huge loads on their heads, walking up past me, smiling with my greetings in Creole, "Ki gen ou Ye", and responding, "Pa pli mal, non, grace a Dieu (Not worse, no, thanks to God)". One might think this remark due to the ravages of the earthquake. But no, it was how Haitians always have responded to each other, something I noticed 50 years ago, reflecting their humble reverential deferential philosophy of life. I smiled at their polite responses, sneaked a bite from a Power Bar, and went back to work.

Our last patient, a 15 year old wearing a beautiful floral print dress but with a horrendous, facial burn (now some 3 years old), was brought in by her mother because she frequently peed on herself.

It turned out she did this only during seizures, and was fastidious and immaculate otherwise. After complimenting her on her lovely dress, we said the bad news was that it was normal to involuntarily lose urine during her generalized seizures. Mother and she and teasing friends needed some good old fashioned medical counseling about this. But the good news was Dr. Judson and I felt her seizures could be brought under much better control so she wouldn't suffer this additional indignity so often. I also took a picture of both sides of her face, showing her the beauty of one side.

I said I would consult a burn specialist friend of mine to see if anything might be done about the other side, which I felt would improve anyway over time, as keloid formation slowly subsided with help from the emollient ointment we gave her. I was grateful at this moment for my 5 years working as the consulting psychiatrist on the Children's Hospital burn unit. I intended to consult my buddy, Dr. Fred Stoddard, Chief of Psychiatry at the amazing Boston Shriner's Burn Hospital in Boston to see what he thought. We needed to determine if they could help this courageous Haitian teenager.

Friday, March 19, Chez Les Soeur: Folding My Tent

Once again I'm going to give you a taste of the guts of our work, taking you back into the trenches with me, where the loves and dreams of my Haitian patients are buried, looking for signs of life and hope, and a way up into fresh air. I want to give you a sense of what it was like working through a clinic morning. After this, I plan to give you only the more interesting cases. I must admit, though, that all my patients, when I listened carefully, were interesting. And they were always much on my mind, often weighing heavily on my heart. It felt like I had just arrived, yet I knew I would be leaving soon, and would have to suffer wondering how my doctors and our patients were doing after I left. Even as earthquake time seemed to stand still for Haiti, so trapped in her disaster, my time seemed to be flying by. I was suffering time warp.

Nestled in the midst of another tent city, opposite the large Convent, our Chez Les Soeur mobile medical clinic was housed in an olive drab army tent. We set up shop in the back and greeted Dr. Marie-Carmelle Louisjas, our Haitian doctor for the morning. Our first patient was a handsome, smiling 14-year-old, brought in by his mother. She complained that he incessantly sang the same song over and over again, lost in thought, oblivious to their attempts to stop him. It was really getting on everyone's nerves. He also had the bad habit of hitting his head against the wall, occasionally muttering he liked hurting himself, saying, "My head doesn't work right." This had been going on for years, but got worse after the quake, when he got terribly frightened and didn't sleep for three nights. When we asked, we found out his grandmother, about the only one who had patience for him now, the person who walked him to school each day, had been killed in the earthquake. He was distraught but had never cried. When did all this begin? Long before the earthquake, when he was nine, and got hit by a car. The impact tore his thigh open and ripped off one of his testicles. He had lost consciousness, awakening a different person.

Formerly a good student, he found himself suddenly slow and inattentive, learning with difficulty. And for a guy, part of his prized equipment was missing. We talked with him and his mother about the combined impact of his brain injury and physical trauma being a huge blow to his pride, making him beat his head against the wall in continuing frustration. He knew his head wasn't working right, and his ensuing bad behavior had worn out his welcome except with his grandmother, who was now dead. His mother began to cry and hold his hand, as tears ran down both their faces. Late in the interview we found out he was already seeing a psychologist, but felt the treatment wasn't helping his bothersome singing. I said I thought he was trying to drowned out his sorrows and console himself, and keep himself company, since there was no school because of the quake and no grandmother to love him.

I tried some of this in Creole, and Tessier and Marie-Carmelle helped me get through. I so wanted to be in intimate personal touch with these patients, and found it terribly frustrating to have the language barrier and need for translation. Tessier and I talked about this, so he wasn't irritated by my awkward efforts to communicate. And slowly things got better. The rust was coming off my speaking ability. But understanding lagged behind. Late in the interview I also mentioned methylphenidate (Ritalin) might be of use sometime. I couldn't resist asking mother, as she was leaving, because of her last name of Calixte, whether she was related to the family I had lived with in Brache, outside of Leogane,. No such luck. My reaching for this unlikely possibility revealed just how desperately I wanted to be in touch with them. I wanted to find out exactly who survived, who got hurt, and what happened to their homes. I longed to help them directly. Had Ternvil's house survived? Had my own room collapsed?

Our next patient was a 9-year-old boy who had a bad infection when he was three and developed a seizure disorder. He had the kind of seizures set off by sleep onset, and was having up to three a night. He was becoming afraid to go to bed, and his mother shared this fear. She turned to me and asked if we could make him normal. Dr. Marie-Carmelle looked at me, and I whispered, "Not normal, but much better." I felt his seizures were out of control. It turned out he had only been on Phenobarbital, and had lost all his medication in the earthquake anyway. We put him on Carbamazepine (Tegretol), and the next week all his seizures had gone away—until the night before our follow-up visit. On questioning, it turned out he had developed a high fever just before sleep, something that he had been getting for several days now, not just at bedtime. I worried about a medication side effect, but Marie-Carmelle knew better. He had been bitten by a mosquito and had gotten Malaria. You think I would remember that pattern better. But she did, and gave him Chloroquine. The next week he was fever- and seizure-free. Through all of this he never complained at all, until that last time, when he said his head itched. Marie-Carmelle examined him, and found he had Tinea Capitis, a fungal scalp infection like ring worm.

Our third patient was a 32-year-old woman with periodic headaches, over-breathing, and palpitations, beginning after her father's death a few years earlier, and now much worse since the earthquake. She had lost her house, was living in a makeshift shelter with her mother and husband, all worrying the rainy season was coming.

"Any other losses?" we asked.

"Yes, my best friend, my only friend, died—I feel like killing myself at times."

"More because of guilt or the pain of missing her?"

"I miss her terribly."

"Would she want you to do it? Would it do her any good now?"

 "No." "Oh, and my scalp, it feels like something's crawling in it." I thought 'conversion reaction, then thought better of it, asking Marie-Carmelle to check. She had Tinea Capitis too, so we gave her anti-fungal cream. She also complained about her inner thighs, suddenly spreading her legs to show us.

"Scabies," said Marie-Carmelle, and another cream was added to her list.

"And I can't read and write well, like others in my family."

"That's inherited", we said, "And you're in the company of your family on that--nothing we can do. But you know, you're beyond school, so you already know you have to live with that. But you do have an over-reactive stress and fear response, giving you the rapid breathing and bursts of fast heartbeat."
Then we took her blood pressure, and it was high, so we put her on Atenolol. A longer-acting Beta Blocker, it would help with the anxiety-related over-breathing and the tension headaches, and lower her blood pressure, as well as other techniques we showed her. We also felt her heart was telling her how much she was missing her dear friend, and THAT reawakened the loss of her father. She bobbed her head "yes".

I looked her in the eye, and said, "Your father and your friend wouldn't want you to kill herself, and now that you have some help, and things you can do for yourself to break the stress cycle, we wonder if you can let yourself go on living?" We were relieved when she shook her head 'yes' again and smiled a little. It was slight, but seemed genuine.

"Can you come by next week so we can check on your progress?" The smile and headshake got bigger.

After standing up to stretch and reduce fanny fatigue, always a problem even with my 'tush cush', I had to have a little talk with Marie-Carmelle. Though doing really well at key points, she was constantly interrupting her interviews to take cell phone calls, and was even doing text messaging (SMS) in her lap while her patient was answering her questions. It was disrespectful, disruptive and annoying, despite her feminine capacity for multi-tasking. I had already told her to put the phone aside, and she hadn't. I asked whether she had a crisis going on, and she just smiled enigmatically. I decided it was probably more in the realm of boy trouble. She was an attractive young woman. I said I'd confiscate the cell phone from her if she did it any more. To be honest, she did, and I didn't have the heart. Others might have.

Our next patient had insomnia, shortness of breath, palpitations, and headaches. She also heard voices, often feeling like running away. I was 'relieved' when they proved to be the voices of dead friends because that meant she wasn't crazy, but struggling with loss. When we asked about nightmares, she said she couldn't stand dreaming about these friends--they kept coming into her mind, night and day. She had felt like killing herself once before in 2006 when she had troubles before. But now it felt like fire in her head when she tried to think, so she avoided thinking. Sometimes she even saw these friends in the dark of night. Marie-Carmelle was heading toward a diagnosis of psychosis, until I mentioned I myself had suffered some of these things after my father's death, and thought that it was complicated painful, partially arrested grieving. So she began to explore further, and as she did, tears began to flow down her patient's cheeks, and even the sister accompanying her cried too. They were no longer alone but joining together in a community of grief.

We felt the impasse had loosened up enough to allow us to shift, and give her specific tools for her physical symptoms, like the sac re-breathing technique, the Valsalva maneuver, to help her with her hyperventilation and palpitations. To move her worries away from bedtime, since they reared their painful heads as she let down her defenses to try and go to sleep, we gave her some homework around talking each night with families members, long before going to bed. She could coax her thoughts and then the sadness out into the open, so they could share it together, giving these friends their due, and allowing her to fall asleep more easily, having already stirred and then drained her tearful reservoir of sad worries earlier in the evening. I felt stingy about not giving medication, but we felt she didn't reach that threshold and our supplies were short. She came back a week later feeling better. So did I, given a difficult judgment call. But throwing meds at everything was not the answer. Careful listening, followed by thoughtful psychological counseling and first aid techniques can help considerably, counting on mind, body and family to restore healthy responses.

I took a moment out for a deep breath myself, and something caught by ear. I had only been dimly aware of it while so focused on patients. Something beautiful was going on. In the Convent across the street, there was a wonderful mixed secular and religious community event going on, with beautiful choral and solo singing. It came wafting across the mud street, through the banana trees and tents to our ears. Older voices were followed by teens and even a children's chorus. And the drumming, what drumming! I found a joyous feeling sweeping over me. How uplifting to hear a spirited community celebration, with the laughter of children and grown-ups sharing a spirited moment. The murmuring of an approving crowd, the waves of applause, all filtering into our tent mixing with the tears and sadness we were experiencing in the Clinic. We were so close to the belly of the beast even as spirits soared. As I've said before, listening with an open heart and mind is like performing surgery without benefit of anesthesia. The healing comes from discovering and sharing fear, pain and sadness, followed by a thoughtful daring spoken response, given judiciously, sometimes coupled with medicine and stress reduction techniques. It takes 'daring' because we have to say things in vulnerable deadlocked areas, guarded by anger and agony, where angels fear to tread. Doctors often stop dead in their tracks, words getting stuck in their mouths. The singing and music from across the street were deeply healing, salve for my sorely taxed ears and heavy heart at the moment.

Our last patient came with a different kind of story, representing deeper mental problems. She had sustained severe losses of family, friends and shelter, and was profoundly depressed. She also had a past history of significant depression. Her voices kept telling her, "Don't kill yourself, don't do it!", which meant that thoughts and impulses of suicide was threatening from deep within her mind, barely restrained and disguised by no's and don'ts. She also spoke of significant anxiety, jumpiness, and obsessive thoughts. We put her on Fluoxetine (Prosac), warning her and her sister about the added risk of suicide from the initial energizing effect before mood improvement set in. We carefully placed her in the constant hands of her strong family, and had her come back for weekly checkups.

I told Marie-Carmelle that since she was so proficient with her cell phone, she should give this risky patient and her family her cell phone number, should there be a serious turn for the worst. "But Dr. Kent, I never give patients my number. We're not supposed to and I wouldn't like it. My private life is my own and not for patients."

"Not from what I've seen today, Marie-Carmelle. All day today, you've already been mixing the two more than you should, so we will rebalance the ledger in favor of your patients. And anyway this is the exception that proves the rule. You wouldn't want a dead patient on your hands."

"Okay, Dr. Kent. Do you do this?"

"Absolutely, when the life of someone I'm caring for is at stake. And I'll tell you one other thing, keeping the phone lifeline open will give her and her parents great support, probably making it unnecessary to call you and giving them a constant infusion of supportive caring. It also makes your cell phone a tax deduction, though probably not relevant here." We didn't talk with the patient about her unbearable losses and sadness to any significant degree, waiting until the second follow-up when she was doing much better. It was safer then because she was more ready to deal with these harsh realities. We didn't want to fan the white-hot flames of her conscience at first given the imminent risk of suicide.

We picked up our chairs as the two doctors and nurses packed up their meds, bandages, and instruments. Together we filed our way out through the tents and piled into the waiting Nisan patrol car—the driver always required to wait nearby for security reasons. I was hot and sweaty, my shirt mottled with wet spots edged by the white of dried salt. I recalled how 50 years ago I would walk over a mile to the 'Clise' when so hot and sweaty, where a huge old iron molasses cauldron had been placed at the crystal outflow of a cold spring. It was a rare luxury to jump in, submerge my salty head, and be churned in its hydraulics until cool as a cucumber. After my pressure cooker under the Trois Soeur Clinic tent, it was refreshing to step into that air-conditioned Patrol cab, and sit in peaceful silence, or maybe listen to some Haitian Merigue or Badou, cued up by our cool driver. I often passed the time with the drivers, asking about Haitian pop culture or family doings, my creative Creole tickling their funny bone. Our drivers rotated among us, but I got to know them all pretty well. Baron Samedi, the dispatcher, remained my favorite. And our rapport would come in handy the next week. I was getting the feeling the gods were still watching over me, even as the frustration of tight security seemed to be further constricting my aching heart. How were my friends doing in Brache and Masson, right at the epicenter of the quake?

Stephanie had announced we were finally to move from the Office to our new Residence that afternoon. Several of us grabbed our stuff, loaded it into a Patrol car, and headed on over. My Shelter Box tent was struck down unceremoniously right after I left. I never looked back—until the next morning at the Office, when I saw it flattened in a heap on the ground. Why had I been so enamored of her? Given the crescendo of activities from this point on I had precious little time or mental space for all the primordial ruminations that had beset me in that tent. Right away my life became more complicated. For starters, on the trip over to the Residence I lost my progressive glasses. God knows how. Though I looked high and low I never found them. Luckily I had stuck an older pair in my suitcase at the last second. Thank goodness. When I opened the case they turned out to be my tennis glasses, with no prescription for reading. I was blind as a bat at close range. How was I going to do my work? I was a bit panicked, but first had to settle into my new digs.

Because keeping the volunteer doctors at the Royal Hotel cost too much, Stephanie consolidated us into this new house, christened The Residence, just a few blocks away from The Office. Crystal, the cook, went with us. On first blush,

The Residence looked impressive, a large, two-story concrete house with portico, surrounded by high walls topped with broken glass and barbed wire—a palace by Haitian standards. But appearances can be deceiving. The recently delivered generator was not connected, and never would be during my tenure. We were at the mercy of the local electric company, which only supplied electricity from 6 pm to 6 am, with random blackouts thrown in for good measure. Though the fridge was good, it didn't work well without electricity. The food got ripe during the day. And my long-awaited bottle of coke was always tepid.

Plus there were no light bulbs anywhere, just those puny power saver fluorescent curly-cues Patti hates. And would you believe, only the 14-watt variety. I have come to loath them myself. When the sun goes down, it is impossible to see, but half the sockets and switches didn't work anyway. The smart young medical volunteers had done what they were told to do in the IMC instruction letter, and brought along strap-on forehead flashlights. Everyone walked around looking like Cyclops illuminated, except for Jattu, Tom, and me, hulking old dinosaurs walking around in the spooky shadows. We went incognito except when we bumped into each other. And there was no Internet, only back at the Office, and that very untrustworthy. My initial enthusiasm was fast fading.

The place did have showers and bathtubs. But there was a little problem with the water supply. The drinking water came in huge bottles nearly requiring two people to tilt them for a glass. Back at the Office we had a tilt stand making it easy, but not at the Residence. During my time one never appeared. I finally felt like stealing the Office stand to stimulate Stephanie to do something about it. I knew how to deal with administrators. But I refrained. Never mind. We had a zillion sealed plastic bottles of sanitary drinking water. IMC had received umpteen palate loads of them, filling every nook and cranny. One of our smart docs finally read their labels, noticing that they were actually mineral water, and warned that if we began to get loose bowels, that was why. So much for good hydration and fluid retention. Actually, they didn't bother me.

But at least we had showers. Or did we? I never saw more than a paltry trickle. And that was when the house water system occasionally coughed up any pressure at all. Whenever the roof reservoir was filled to the brim, you knew it because it overflowed onto the roof and cascaded down over my window where I hung my clothes to dry. Somebody would finally yell about the overflow cascade and the security guard would drop his shotgun and come running around the back of the house to turn it off. Of course, we had the same high tech way of finding out when the roof reservoir was empty. Suddenly no running water. So the showers dribbled cold water at best. Note I said cold, meaning no hot water at all. When the reservoir was first filled, it actually was quite bracing, but soon, with the sweltering heat of the noonday sun on the roof, it would warm up to tepid, kind of like my coke.

I should also mention that our downstairs bathroom had no working light socket. We peed by candlelight. Very romantic, though damp at times. And the handle came off the bathroom door from the inside, locking us in but no one out. We got to know each other very well. The indoor gecko population was hard working and kept the mosquitoes down. It would have made a good Geico ad. But these cute little creatures were no match for the two-inch spiders that lurked or crawled about on our ceilings.

Then there were the ants. They were extremely tiny, but incredibly fast, and came in hordes. We didn't need to sweep the floor. The ant army carried all droppings away almost as soon as they hit the floor. They also seemed to enjoy my computer keyboard. I must be a sweet typer. I probably should have washed my hands more often. Once I forgot that I had left candy bar crumbs in my backpack. Putting it on the next day I soon realized half the anthill must have been inside munching away on Power Bar crumbs. Suddenly they were crying Mayday, because they came pouring out all over me. Though little they packed a big bite. Medically, one would diagnose me as having a bad case of formication—an actual bona fide diagnosis (usually reserved for a drug side effect).

 One night, a medical volunteer was horrified when she saw me preparing to do carnage to these ants. By accident I had dribbled some coke on the floor and the ants discovered my dribblings. Soon a horde was feasting at the edge of Sugar Lake. When she wasn't looking I poured beer all around them and watched it flood them until they drowned. Sorry you right to lifers. I am usually on the side of good and right, but something was pent up in me from my day's work, eating away at me. All day long the sad chaos of the earthquake trauma I had to fight in the ravaged minds of my patients flooded in on me, often making me feel helpless and at a loss for words. With each new case, I would initially go through anxious uncertainty about finding some useful diagnosis and helpful words to say. Playing all-powerful god for a second against the forces of evil ants felt relieving. Shrinks get to play a little too, sometimes. When Stephanie walked by she put her foot in the middle of my handiwork before I could warn her. Truth be told, we all fought the ant wars. Earthquakes and mass human misery reduce us all to feeling puny and helpless.

But with topflight volunteer docs around, the attack on imagined rampant Haitian microbes was even more relentless. The medical doctors among us were obsessed with contamination, and brought along bottles of Clorox to disinfect the dishwater. You see, I wasn't the only paranoid around. They also used it for other unmentionable things. When Crystal clogged the kitchen sink, I dismantled it, successfully setting loose a torrent of dirty dish water all over me and the floor. One of the doctors came rushing over with a big bowl, to catch the torrent, forgetting the bowl was already filled with Clorox dishwater, dumping it on my head. At least the drain was clear, and the floor clean, plus my hair and shirt didn't need washing for the next two days.

Ruth washed all our clothes, but then piled them in one huge hamper for us to hunt through. Soon I found myself down to two shirts and two underpants. I wondered if I would see them recycled at the Haitian resale stand on the street outside our house. At least the artwork in the house was uplifting, especially the central painting, 8 feet long, and 4 feet high, of a naked slave chained to a wagon wheel, silhouetted by a neon sunset. Though reflecting Haiti's past, it was so depressing in the dim fluorescent lighting we hid it away for the duration.

Being a 'one-month'er' I qualified for inside bedding. Others were outside in some tents. Being inside would be good, especially if the rainy season descended, unless a strong earthquake hit, making the tent safer. But the fact that this house withstood the first two major 7.1 and 6.4 shakes was a good sign. I found I was to bunk with Tom, a Kenyan latrine specialist. He was a large dignified, imposing man with a deep resonant voice. Luckily, he didn't snore. He preferred to drink countless Heinekens, and creep to bed long after me. I left the light on for him, protected by my eyeshades. In my bed I had to be very careful. It had slats to support the mattress, and they didn't reach all the way across to the other side. The uncomfortable middle support from foot to headboard allowed the mattress to pivot if I dared cross the midline, dumping me under the bed, my own version of an earthquake nightmare. I learned to lie on just one side of the bed, carefully staying on my back, each night lifting Norbert's accordion-framed mosquito netting up over me, from my feet to my head, a bizarre sleeping chrysalis as I lay their enshrined in my nylon running shorts, Obama staring up from my T-shirt. I must have looked a sight for Tom each night after all that beer.

Tom took umbrage at this passage about him being a latrine specialist when I read it to him, reminding me he was really a Water and Sanitation Specialist. "Kent, can you imagine trying to provide water for tent cities of up to 40,000 people; and even more, can you conjure up how much waste, and how many latrines are required. I am holding dehydration and disease at bay single-handedly, like the boy with his finger in the dike, standing there between them and a rampant Cholera epidemic." Once he painted this graphic picture, I stood in awe of Tom, protector of the vulnerable and destitute masses. I put my hands up in respectful surrender. "But that's not the hard part,' he said. "We have the technology and the equipment. It's dealing with the government, and the competing interest groups, that's the difficult part. I have to wade through a labyrinth of bureaucratic cesspools. Pardon my metaphor."

Tom was a great raconteur, and veteran of many crisis situations—Somali, Chad, Darfur, Ethiopia, Congo, and now Haiti. He seemed to enjoy telling me war stories: for instance, the one about Darfur and rebels crawling over his compound's barbed wall, hot wiring his vehicles in the dead of night and screeching out of the gate. He was furious. They had awakened him from a good night's sleep. He particularly enjoyed telling me the sequel, when two contingents of the same group, unaware of each other, were attacking the compound at the same time and encountered each other not far outside. They opened fire on each other, thinking they were being ambushed, and decimated their ranks. Only the bloody spots on the ground gave the body count. The remains had been hauled away by rebels, or animals. He also mentioned the Ethiopian famine, when he watched a starving mother, as food was being delivered at the other end of the square by an NGO. She left her baby asleep on a mat to run and get their share, way across the square, not realizing her child awakened and came crawling after her. Suddenly a black vulture came swooping down and grabbed the baby in its hungry talons, and made off with it-- to Tom's horror. The mother never realized what happened but went crazy over her missing baby. As the Heinekens flowed, the beer helped his story telling, but not my subsequent sleep.

I didn't know if Jattu approved of her fellow African drinking beer, though I must tell you this dignified, jovial Kenyan was never even tipsy. He knew how to pace himself. Jattu, my Russian- and French-speaking doctor from Sierra Leone, was always dressed beautifully in bright colored traditional African fabrics. Yet a smile rarely brightened her face. Always stately and alert, Jattu was deadpan. Yet her voice was lovely, and she always prefaced her most intimate comments with a disarmingly formal, "Dr. Kent." "Dr. Kent, why don't you take better care of yourself?" "Dr. Kent, you forgot to comb your hair!" " Dr. Kent, why don't you take more water with you?" She lived directly across from me, with that first floor bathroom between us. I felt like I had acquired a lovely new African mother, although she had a slightly different look in her eye at times. I thought her Muumuu-like dress was to conceal worries about her slightly ample figure, until she confided in me, "Dr. Kent, I'm really worried about something. Because of Crystal's cooking, I'm really losing weight, and African men don't like their women skinny."

This comment made me think maybe Tom was winning our Jattu competition, until one evening I overheard her saying, "You know, I saw Dr. Kent doing his exercises this morning. He really has quite a routine, and some of them are exotic." *Did she say exotic or erotic? I wasn't quite sure.* I had no idea she had seen me doing my weird calisthenics. Doing them in my tent was hassle enough, but finding a private place in our crowded house was a real problem. I had tucked myself away in our tight little room, only my feet sticking out. But she must have spied me from afar. The next thing I knew, she announced, "Dr. Kent, I'm going to get up and do them with you." I relish my private time in the mornings, and suddenly felt invaded.

"Jattu, they're not easy, and I do them at 5:30 in the morning."

"I know, Dr. Kent, I'll be there."

And she was. She had shorts and a top on, and plopped her Thermorest camping pad right down next to mine. Begrudgingly, I began explaining my arcane sequence, demonstrating each one as I did my drill. She copied along, stopping about halfway through each one, but doing them all. And the next morning there she was again. Apparently no random flesh in the pan. By the third morning, I had had it. The fourth morning I needed to get up really early anyway (to prepare my next Saturday Training Seminar), and popped out of bed at 5 am, throwing her off my tail. When she came out at 5:30, mat in hand, she pouted when she saw me sitting in front of my computer, already dripping wet. "Dr. Kent, you didn't wait for me!" I felt like a heel. But I had shaken free from this designing African woman--probably all a big fantasy in my head. Imaginary intrigue makes life more exciting.

Jattu was a physician, brought aboard to oversee the clinics and the pharmacy. At one point she noticed one of the clinics was getting quantities of injectable Demerol, a very addictive black market drug. She put a stop to it, though we never found out what it was being used for, or, whether it was being sold. She was a smart cookie and nothing ever got by her—except for Crystal, our cook. In exasperation over inadequate quantities of food for dinner, and repetitive menus all the time, we finally held a little town meeting with Stephanie. Tom got on his soapbox and gave a remarkably articulate oration about Crystal's shortcomings, spurred by his prodigious appetite. At the end, Jattu made an impassioned plea for a clear approach to healthy adequate menu planning. I threw in my two cents worth, first complaining that I had been asking for scrambled eggs for over a week, and she had yet to deliver them before I left, occasionally putting them on the table just as I was leaving, just to torture me.

My second cent's worth was that Stephanie, who was always overburdened, should consider delegating all this to Jattu. For once, Stephanie, who was a strong leader, caved in, and gave the job to Jattu. The next morning she had a little meeting with Crystal relaying all this, throwing in for good measure that she had to arrive at the residence by 6 am so Dr. Kent could get his eggs on time. You should have seen her scowl when she arrived at the Residence the next morning. It was 6:30 am again on the dot, not a second before, and my eggs arrived just as I was leaving for the Office. And this is not the end of the story.

What I'm getting at is that, despite the shortcomings of our Residence, (certainly vastly better than a make-shift Haitian lean-to shelter), we found its creature comforts challenging, yet it became our home, especially with our unique group of people aboard. In that number was dear Dr. Alice, who lived not far from Tenant's Harbor, Maine, where I summered as a youth. Alice, a volunteer family practitioner, had decided to stay on with us because she loved the clinics. And she became my heroine. When she heard my spectacle crisis, she beckoned me over and reached in her bag. Out came an extra pair of reading glasses she had brought along. She was a lifesaver. And I was back in business!

Dr. Paul, from Rush, and a nurse practitioner Alisia, from Alaska, arrived Friday night, and settled in over the weekend. Paul and Alisia slept upstairs. The next morning, because it was a Saturday, we all planned to sleep in, no clinics to rush off to, no seminar to do. But this was not to be. A little before 5 a.m., when the chill was still in the air and the first rosy fingers of dawn were barely spreading across the sky, a god-awful racket pierced the air. Someone was playing loud rap music outside, someone else banging boards. It came ricocheting down from the roof of a huge new construction project just outside our window, just beyond our glass-studded, barb-wired wall. Security didn't keep the sound out. Unable to sleep, my curiosity peaked; I crawled to the window, parted our tattered gossamer drapes, and squinted up. There silhouetted on a steep roof were three figures, one guy jiving to a boom box, the other jabbering away in stentorian Creole, and a third actually working. They were perched at a dangerous angle, apparently getting the roof ready for an afternoon pour of roof cement. Even Tom, a prodigious, well-sauced sleeper, groaned and rolled over, smashing pillows against his ears. Pretty soon these early birds were joined by a large crew of roofers. Tom and I gave up. We were off to a fine start for the day.

Alice's glasses came in handy Saturday, despite my yawns. I had a lot of seminar prep to do. Uh, oh, there was that old familiar sinking feeling in the pit of my stomach. But I now had a week's experience with the Haitian doctors under my belt and knew what they needed: the 'low down' on how to do the psychiatric and mental status exam, based on the cardinal symptoms of major mental illness. I spent the day preparing everything, but then had to translate it into French and Creole. The nurses spoke mainly Creole. My ace translator, Tessier, helped me with the Creole the following week.

But what to do Sunday, my only real day off during the entire month? While I was having dinner with Tom and others Saturday night at the Hotel a guy from the Royal came around offering a trip to an undisclosed beach, with a great lunch and lots to drink. Tom and I had already had our beers and Cuba Libras, and we were deep into Strict Badou, the Hotel's throbbing melodic island band. So we called up Jattu and the docs, and made an executive decision. We were all going. The only problem was Stephanie. And security. We figured a *fait accompli* was the best approach. Though a wounded worried warrior, Stephanie was a good soul, and being French, well, Canadian French, she understood enjoying life, though she worked herself to the bone. When we handed her the fact that we had signed up, she momentarily went conservative, saying it might not be safe, and we probably shouldn't go, not unless we could get the name of the beach and find out if it had cell phone coverage. "You never know when trouble might strike," she said cautiously. When I told her Hugo was in charge, the Royal owner she did business with around the Seminars and the volunteer lodging, she lightened up a bit, saying, "Well, okay, but you're going to have to take the satellite phone; I'm charging it up as we speak."

I went to sleep with a smile on my face. We had a plan, and the roofers wouldn't be working on a Sunday morning.

Sunday, March 21: *Replique* (Aftershock)

I sat bolt upright in the dead of night, awakened by something—and from a delicious dream, too. My roommate, Tom, let out a loud grunt, and rolled violently across his bed, landing with a thud on the floor. He lay there deathly still. I was about to check him out when he leapt up and ran out of the room. *Must have been one hell of a nightmare*, I thought. What confused me was the rumble of low talk outside. I could see Jattu's light on, but that wasn't unusual. She didn't sleep well and often got early morning calls from Sierra Leone. Then I thought, they're *partying out there, finishing off my new bottle of Barbancourt I left on the table. Damn!* But I was too tired to join them, and drifted off. It was only 2:15 in the morning.

When I sat down at breakfast I told Jattu about Tom's weird behavior. She said, "Dr. Kent, that was no nightmare. That was a '*replique*.' He felt the aftershock last night and headed for the floor. Most of us woke and ran outside. Where were you?"

"Aftershock? I didn't feel anything. Maybe that's what woke me."

"Stephanie got out too, and Dr. Paul. They all felt it. You didn't hear the rumble, and feel the shaking?"

Dr. Paul cut in, "It wasn't a rumble. It was a cracking boom, and then a scary side-to-side shaking. I ran out, but then felt I wasn't being chivalrous, so I ran back in, and upstairs, to get Alisia. I found her standing there looking quite dazed. She wasn't worried and nothing more happened, so we went to bed."

"And Kent, nobody touched your precious Barbancourt," said Tom. "But I thought about it. Makes earthquakes go down easier."

"Damn! I missed the whole thing again! That makes two in a row. Well, to hell with it. At least we've got a day at the beach to look forward to."

Later that morning, Stephanie called me over to the Office. Oops! Was I in trouble for instigating the beach rebellion? Turns out she wanted to put me in charge of the satellite phone. Maybe she had glanced at my novel **Body Sharing** floating around the Residence. If so, she might think I really knew something about them. She walked over to me, and I suddenly found myself cradling the handset in my hot little hands. I couldn't believe it. I was actually holding an Iridium Satellite phone.

You see, Iridium placed 66 satellites in low earth orbit a few decades ago at a cost of about $5 billion, for real time mobile phone service all over the world. Iridium went bust, but the satellite system was purchased by a private equity group for less than $50 million because the satellites were projected to fall out of low orbit soon. The mobile phone service was used principally by the U.S. Defense Department for troops and field commanders needing real-time ground communication. The purchase proved propitious, since the satellites exceeded by two fold their nominal design lifetime of five years, creating a significant financial return for the purchaser.

My buddy, Stephen Day, painter, writer and raconteur extraordinaire, introduced me to the General Council of the original Iridium venture. We all belong to the exclusive Friends Creek (trout) Fishing Club. A little later, I ended up treating the son of the woman running their legal department. And now here I was actually holding one of their phones. Such a small world. I was raring to try it out. This was the very phone I have Rex, the protagonist in my novel, **Body Sharing,** using, when he gets attacked by a Grizzly Bear. Star Wars to the rescue.

Up to this point, we had no idea which beach we were going to, though Hugo thought there was cell phone coverage. But no one was going to pry this Iridium baby out of my hands. We headed off for the boat. And, surprise of surprises, we went to the same launch place used for our Platon Boat Clinic. I had a hunch where we where going-- straight toward Platon.

As our two boats skimmed along toward our noontime feast, there was an irony to our boat ride. Emblazoned on the side of each craft was the slogan, 'Food for the Poor." Of course, this group raises great money to feed the hungry, but I had to reflect once again on how fortunate we were, laden with our feast, even if it was our only real outing in a month.

We stopped short of Platon, turning into Bananier, that superb little crescent shaped beach we had seen, with a few huts, lots of coconut trees, and a big tin-roofed, open-sided cook area. Oh, and, of course, abundant Banana trees, growing in tiers behind the towering coconut palms. We splashed ashore while the staff ferried precious cargo onto the beach. Jattu, Alice and I spread our purloined sheets around the base of a big Papaya tree, while Tom, no Heinekens available, popped a Prestige Beer.

He never carried a bottle opener and had this uncanny way of using one bottle to open the other, making a resounding, champagne-like pop. We were impressed, until Jattu opened hers with her bare teeth. I wasn't about to try this, having just spent $38,000 on 15 crowns--my teeth worn to nubbins from listening to my patients, tooth grinding, and the fact that I drank nothing but coca cola all day long. The latter was my wife's diagnosis, but my father had the same tooth problem, and he never drank coke, or treated patients. Old wife's tales bight the dust. Genetics reigns.

I flipped a nearby cooler open, and grabbed The Real Thing. Tom offered to open it for me. I preferred his pop to letting Jattu shatter a tooth on my behalf. I would be too indebted to her. The sun was hot, the shade cool, and the gentle breeze refreshing, as we contemplated our next leisurely move. Jattu entertained some kids while we discussed the matter. Alice wanted some coconut milk and Tom, handy with a machete, cracked one for her.

Jattu entertains kids above and Tom cracks coconut for Alice below

Dr. Alice ripped open her Velcro pack pocket and pulled out miniature binoculars. She was a closet Birder. Actually, she was a top ornithologist. At every turn, she amazed me—until it came to washing dishes one night. I was using tap water from our trusty roof cistern, admittedly cold, but very soapy. I was halfway through rinsing when she accosted me, "Is that water disinfected?"

"What do you mean, I brush my teeth with it!"

"You have to be kidding. God knows what you'll come down with. Here, use this Clorox, or we'll all get dysentery."

I was indignant, but when the rest of the volunteer docs ganged up on me, I vacated my post, letting the Clorox Queen take over.

"Hey, Kent, come, look at this!" Alice yelled. "See that Tern in the hole over there in that big palm. I'm sure it's a first for me." She was flipping madly through her trusty bird book. I had to hand it to her. She was as adept and complete at birding as she was at medicine. Jattu was asleep, so I headed off to sit with the locals at a table in the back palm grove.

After a weary silence, I muttered a few tentative words in Creole and they began chatting with me. I was happy to respond, because when at Platon, I was all business, no time for small talk. I wanted to find out about these remote fishing communities, so isolated with no road to civilization. How had they done with the earthquake? One guy's eyes got big. A buddy jabbed him with his elbow.

"Okay, I was standing on the shore down there. My brother was out with his dugout , diving for *lambi*. He was under water. The next thing I knew the ocean started to rush out away from me, scaring the hell out of me. I looked out to sea in a panic, seeing my brother's boat spinning out with the receding water. A minute later a huge wave came roaring back in, carrying his dugout, but he was nowhere to be seen. We found his body the next day."

Tears began streaming down his face. A buddy put his arm around him. I put my hand on his. I told him about what had happened to some of my friends in Brache where I used to live. There is something about grieving that cuts through country and color. He wasn't the only one at sea, or on land, that suffered casualties from the quake. I recalled that in Petit Guinee we had driven past part of the shore, dead sunken trees witness to a vast swath of beach that had caved in, the salt water creating these lonely skeleton palms. Houses had crumbled, but now truckloads of other destroyed houses were being used to build it back up. I kept these thoughts to myself.

I wondered if they would do a '*retirer en bas de lo' ceremony*' for the dead? It was my way of checking out whether there was a local *Houngan* practicing voodoo nearby, since in Brache these sad losses would be mourned and handled by such a respectful voodoo ceremony—though technically not until a year later. They were cagey at first, until I let them know about living with a priest some time ago. They didn't believe it until I threw out a few Voodoo terms and named a few arcane *Loa*, like the Petro god *Simbi en de zo*, and *Maitresse Erzulie Gran Freda*.

Finally, I said, "Look, you guys are fishermen, so you must have a ceremony for *Agwe Ta Woyo*, and *Maitresse La Sirene*, god and goddess of the sea, to insure fisherman safety and prosperity." This touched them where they lived, and an animated conversation began, letting me know there was, in fact, such a 'native healer' up in these hills. I was interested, but the conversation petered out. I finally asked why they were being so careful around the subject of Voodoo.

One fisherman cleared his throat and said, "Look, we were told that the tourists would be curious, but we had to be careful, mainly because some of them, especially the whites, are quite misinformed and superstitious about our belief in ancestor worship. We respect our elders, and when they die, we slowly deify them, trying to collect all their accumulated wisdom."

"Yeh," I said, "and you've probably heard we don't really respect our elders, or their wisdom, so we wouldn't understand."

"And that you wouldn't understand how we could practice both Catholicism AND Voodoo."

"Right," I said. "Given your lives in Haiti, and all you've been through, now including the earthquake, you want to respect and serve all the powers that be, neglecting no one lest there be another catastrophe, and you were betting on the wrong god."

"You do understand our situation," said the guy whose brother had been lost. "And we hear tourists wouldn't understand our ceremonies, thinking they are primitive."

"Oh," I said, "You mean your voodoo animal sacrifices? Well, you have to consider our mature beliefs, like 'eating the body and drinking the blood' of Christ. I had one guy from Africa tease me we must be cannibals." Everyone laughed.

"Yeh," said the fisherman, "We prefer to substitute chickens, goats and bulls for our ritual sacrifices. But on the other hand, we're Christians too. So we know what you're talking about." Another guy chimed in, "Anyway, we're not supposed to freak out the Hotel's tourists with our Voodoo stuff. They're not ready for it and it would be bad for business. We worship the greenback and depend on the income." By this time my hunger was beginning to peak, so I said goodbye.

I wandered over to the cooking shed, chatting up the cooks and inspecting the roasting lobsters, frying chicken, and simmering conch stew. The aromas were tantalizing. A few steaks and pork chops were also grilling, but to a crisp I feared. I asked why.

"To kill any worms and bring out the flavor," said a plump jovial cook. "You guys like them raw, but we find that unsafe and a bit, how do I say it, primitive. But we don't have to worry about that with our lobsters and conch."

As I stood there, watching the smoky vapors rise all around me, the smells were mesmerizing. I watched another cook tending the little charcoal lobster stoves, and saw her reach down and rearrange the coals with her bare fingers. Back when I lived in Brache 50 years ago, Joselia, the Houngan's wife, would do this when she cooked goat meat for me.

"How can you possibly pick up burning coals like that?" I once asked her.

She replied, "You have to be tough and have big calluses, Sparky (my other nickname back then), to tell you the truth, you have to have a good eye, picking up the side that isn't burning. Only the god, *Simbi En De Zo* can really walk on a bed of burning coals, and when the peasant he has possessed wakes up, sometimes his feet are a bit sore."

The pungent smell of *'di ri ak pwo ruge'* (classic rice and red beans) brought me back from my reverie. But what really caught my nostrils were savory fried rounds of *'banan pese'*. My mouth began to water. I hadn't had fried plantain for several days. Given my appreciation for this rich mix of traditional Haitian cooking, I couldn't help but share enthusiastic compliments. Enjoying them, they treated me to a taste of the rice and beans. Excellent! Crystal wasn't bad, but she just didn't get it quite right.

After while, I ambled back to my blanket, bedding down between Alice and Jattu. Alice, back from her birding, had her other pair of reading glasses on and was buried in her bird book, while Jattu was snoozing quietly, reclined in an Odalisque position. I was just drifting off when Alice tapped me on my shoulder. One of the guys who had been chatting with me in the kitchen wanted to deliver a piping hot platter. Standing ceremoniously over me, he offered me the very first lobster of the lunch. Everyone was looking on enviously, trying to figure out why he would seek out some joker lying asleep on his back. But I knew why. And the lobster and fried plantain were delicious. Being first to finish, I began to think about where to go next. And it was into my fanny pack. The Iridium satellite phone. Tom eyed me as I pulled it out.

"You can't be thinking of using that," he said. "Do you know how much each call costs?"

I said, "No, and I don't want to know. Stephanie gave me this in case of emergency, and I have to see how it works. Better now than when the car thieves and vultures begin their attack."

"So who are you going to call, Stephanie?"

"I wouldn't want to bother her. She'd freak out thinking we were being abducted. No, I think I'll just make a little call to Patti in France."

Tom winced, and then smiled. By this time, Jattu was awake, and Alice was examining the gadget over my shoulder. "I'm next!" she said. I'm not shabby at electronics."

She was able to get the thing to light up. I punched in Patti's French cell in the usual fashion and waited. Nothing happened. Not even a buzz. For the next ten minutes I fiddled with it, trying every access number combination I could think of. Nothing, not even a burp. In exasperation, Alice snatched it out of my hands. With finesse, she dialed in. Again, nothing. She worked her magic for another several minutes, and gave up.

"All right, Children, let Papa show you how," said Tom. We all assumed this world traveler, veteran of many abduction-ridden campaigns, was an experienced old hand and would put us to shame. No such luck. We all turned to Jattu, half expecting her to rescue us. But she just shook her head. So we fanny packed the project. So much for Stephanie's fail-safe fool-proof security strategy. The Iridium was stowed away with shared disappointment

The problem was, I had now worked up quite a sweat in the process. I looked out at the sparkling blue Caribbean, reached back into my fanny pack and pulled out my goggles--and my sunscreen, asking Tom to slather my back. A moment later, when I thought no one was looking I dropped 'trow'. One eye cocked, Jattu hadn't missed a thing, and began muttering something about my preferring to have Tom put on sunscreen instead of her. Finally, she pointed out my skimpy tank suit. "Dr. Kent, you really came prepared, didn't you. No need to change if you don't mind wet shorts. And, Dr. Kent, watch out, that sun is scorching. Don't stay out too long."

I looked around to see where Alice was, and spotted her way out in Bananier Bay, swimming alone. I walked down to water's edge, ready to plunge in and swim out to her. One problem. A ten-foot swath of sharp rocks lay between me and comfortable sandy bottom. And I was a tenderfoot. Picking my way daintily through the rock field, I finally hit sand, and water. Stroking out to her, I said, "How could you dare swim out here alone. It's not safe. You know the rule. Always swim with a buddy." Realizing I was giving her some of her own conservative medical grief, she laughed, and said, "Thanks for rescuing me."

She was a good swimmer, and we both caught sight of a lone fisherman and his dugout off to the left by a point. The water was cloudy where we were anyway, no good for skin diving. So we set out. When we got out near him, the water became breathtakingly clear, like back at Il de la Gonave long ago. I was snorkleless now, but breast stroked out to the fisherman. He was after *lambi*, as it turned out. But he was equipped, full-face mask and snorkel. I watched him for a while, and then spun around, fearing I was neglecting my swimming buddy. She was nowhere in sight. I felt like Dr. Paul, having forgotten my chivalry. I looked around frantically, finally spotting her, a speck walking along the beach. She had climbed out on the point, and was making her way back to civilization. And I was alone. 50 years ago I would have thought nothing of it, since I often made *camionettes* stop along the National Route to let me off at god forsaken beaches to skin dive for hours on pristine barrier reefs. Back then, being young and invincible, I kept the notorious Haitian sharks out of mind, pushed way into the back of my preconscious. Luckily, in all my diving in Haiti, I never saw a shark, and no moray eel nipped any of my fingers or other precious parts.

My passion for snorkeling had propelled me into beautiful but dangerous waters. My first diving buddy, Chris Bent, would have been proud of me. He coveted prime coral diving, and was a fearless companion. Back then I thought he would be impressed with my Haiti daring, until a few years later I heard NBC News announce, "Lieutenant Commander Christopher Osborne Bent and his UDT Team are now securing the flotation collar around the Gemini Space Craft, assuring the safe arrival of the astronauts." So much for my prowess, but I was proud of my buddy, now dubbed 'Frogfather' as his Internet moniker.

Okay, so here I was, again in Haiti, swimming all alone, in crystal clear water, with good depth and rich coral heads passing below me. But no snorkel and no flippers. Resolute, I abandoned the dugout, and found the warm current carrying me across the mouth of Bananier Bay, in the right direction toward our beach. To my delight, by just plunging down 6 feet, I found the water turned from warm to pleasantly cool, the sun's shimmering rays only heating the surface. I floated along dreamily, looking for conch (*lambi*), but despite intense scanning found nothing. How was this dugout guy making a living? I have a pretty good underwater eye, and the bottom was picked clean. Then I had a chilling thought. What might be scanning for me? I pivoted to check for sharks and barracuda pursuing me. I used to be more casual about this as a youth, but old fart paranoia was upon me once again. Or was it more seasoned caution? Was I finally growing up or was I chickening out? Anyway, it was only my unconscious following me. The coast was clear. Fifty years ago I used to forget where I was and cruise along half an hour without shark checks. On this dive my latency was a scant 3 minutes. Even so, I was having the time of my life.

After while, I was back in the boat channel stroking right in, past coral heads spaced widely on either side. Almost ashore, I realized I had seen precious few fish. Were they scared of me, or was this place fished out? But I knew they found lots of lobsters. I had just eaten one. Oh, well, maybe they had to go further out. But not me, not any longer. When I climbed out and approached our Papaya tree, obviously an empty handed hunter, I looked around for my erstwhile swimming buddy. She was nowhere to be found, until I caught her off birding again. "Thanks, buddy," I said, "for abandoning me to the sharks."

"Oh, Kent, grow up. You didn't even notice when I yelled I was going ashore."

"Sorry. Next time I'll do better." Soon we were all back in the boats, warm with memories of our day at the beach."

After we got back, I turned the Iridium phone in to Stephanie at the Office before heading over to the Residence, grumbling something about not knowing how to use it. "You don't? Somehow I thought you did. Here, let me show you." She turned it on, flicked a little switch none of us had noticed, and said, "In this mode you can dial it like any phone."

I looked at her, "Thanks for telling me, Steph."

"One other thing. I've been thinking about your wanting to have a meeting with your friends from your field site. As you know, there's a big security problem going on. There's been yet another abduction/ransom situation, and there's a rumor someone from an organization in Cap Haitian was killed. I checked with our top IMC security person, and he said I just can't let you go to your field site, even with armed security."

"Not at all?" Taken by surprise, a wave of sadness hit me, then I started getting angry.

"But I know how important this is to you, and you've really been doing good work, so I have an idea."

"Shoot."

"We've opened a new IMC Office in Leogane, run by my French friend, Alice."

"You mean the Outreach Director? I met her at breakfast awhile back."

"Yes. She's got a new role. Why don't you set up a rendezvous at Alice's place in Leogane? That gets you close, is easy for your friends to find, and nobody upstairs can complain about security. They have a guard there all the time."

"Great idea! Thanks for thinking of me. I'll give my friends a call and set it up." Deeply touched, I realized this compromise solution now seemed quite acceptable, even good in some ways. What was happening to the hair-shirt freedom fighter in me? But there might be a fringe benefit to this particular compromise approach. "I could check out Alice's headquarters, introduce these guys to her. Maybe she could use them in her office." I was surprised I took this so well, but felt quite relieved and grateful.

As we headed to the Residence, I knew another new medical volunteer would be waiting for us, taking the place of departing volunteers. Melissa, the newcomer, hailed from Canada. She had arrived that Sunday afternoon while we were boating back from the beach. Dr. Paul and Alisia had decided not to join us at the beach. Melissa and Dr. Paul were reclining on our Residence portico when we got there.

Paul had worked for several days in the Port-au-Prince hospital before arriving, while Melissa, a new volunteer, was fresh off the plane Friday. She sat there almost mute. Being a shrink, I announced what I did and finally asked her if she were perhaps a bit worried about what she was getting into. "My aren't we projecting, Dr. Ravenscroft," she said, "are you always so nervous when you meet new people?" I knew I had met my match. Though quiet, she proved quite sharp, a doctor's doctor, carrying the PDR and Merck Manuel around in her head. Everyone went to her for drugs and dosages. She also quizzed Tom mercilessly on latrines and water supplies. I listened—for a while—finally finding out what he really did.

Dr. Paul, on the other hand, showed his brilliance and sense of humor in other ways. Later, when I finally managed to get the lab near the hospital to cough up my INR to check if I was properly anti-coagulated, he took one look at the numbers, and said they didn't make sense. The INR was extremely high, meaning my blood was extremely thin and prone to a bleed, so I was relieved he thought the results were off somehow. But then he hedged his bets by telling me not to cut back too much. Waggishly, he said he spent his time nearly killing his child patients at Rush with his anti-cancer medications, but he didn't want to have my death on his hands. He explained that kids had an incredibly high recovery rate from cancer compared to adults, mainly because their parents were willing to have them stay in the hospital most of two years, allowing him safely to use really high doses, something adults wouldn't do. He was totally on top of hematology and oncology, but daunted by tropical medicine. His nose was in a book a lot at first. Truth is, he worried about stuff we hadn't even thought of.

Paul was well equipped with everything, and one night when he was reading on the couch by his head light, we saw a huge spider coming down the wall, about to reach him. Just as we yelled, the lights went out, but not his headlight. All we saw was his light flash past us like a comet into the next room. When the lights came on he had settled into a new chair to read. The good doctor moved faster than the spider and must have been reading something really good. Turned out it was a medical manual, of course.

I was in awe of him, but right off the bat I nearly caused him to have an acute identity crisis. You see, we made it easy for Haitian staff and patients at IMC by calling ourselves by our first names. I was Dr. Kent, as Jattu has taught you. The only problem was, Dr. Paul's last name was Kent, and he was used to being called just that: Dr. Kent. So every time somebody called me, we both answered, and it nearly drove him crazy. But he was flexible, and gave up his name—just for me, and to preserve his sanity. I was there first, anyway, and had secured the high ground. People at Rush will wonder why he doesn't respond to his name any longer when he gets back. He was also rather selfless, giving up his bed upstairs when new staff arrived, pitching his little pup tent downstairs. Where? Right in the front hall. I felt sorry for him sleeping on the concrete floor, so I loaned him my cushy air mattress. I had my lovely bed, anyway.

Dr. Paul seemed so kind and jovial most of the time—that is, until one evening our Nutrition Program nurse got us to play a fast-paced word guessing game. Then his killer instinct came out. He gloated when we stumbled around and the buzzer nailed us, and hooted when he won. But I'd put my life in his hands anytime. He had a heart of gold and a steel trap mind. He brought two big duffels in addition to his suitcase, one filled with toys, the other with soccer balls and sports stuff, all to be distributed at his clinics. He loved kids, like Alisia. Only someone ferrying them from the airport misplaced them, and only the soccer balls showed up. He prayed Haitian kids got all the toys, and not the black market.

Paul and Alisia had a running patter that fascinated me, and kept us all in stitches. Her mother, a bit overweight and irrepressible, worked in a doctor's office, and couldn't help calling up Alisia during working hours as a teen, whispering the choicest morsels to her, like the guy who (prepare yourselves) came in shock, having been examining himself in the shower, feeling "something back there", which he began pulling out like a long ribbon, until he lifted it up to see what it was--a huge Alaskan Kodiak Bear tape worm stared him in the face. He shrieked and came running in to the doctor. Nothing fazed dear Alisia after such careful maternal upbringing. Yet she never lost her sense of humor, and enjoy children, as you will see. Never one to miss an opportunity to teach, Alisia went on to tell us she knew right away from the story of her mother's patient that he was an Alaskan newcomer. All the old timers knew you didn't eat Salmon from the mouth of a certain river because they were all infected with Kodiak tapeworm cysts. From an early age she already had a keen sense of parasite life cycles. From Alaska, apparently you can see not only Russia but the whole fascinating field of medicine laid out before you.

 What does Alisia look like? Not usually vain about herself, one early morning she was primping in front of the big mirror in our entrance hall. I was out of sight doing my exotic exercises. As she combed her hair, seemingly out of nowhere Alisia heard this voice saying, "Mirror, mirror, on the wall, who is the fairest of them all?" Not missing a beat, she said, "Then I'm in trouble."

I replied, "Any girl from Alaska has a beautiful soul."

"Then I'm really in trouble and so is Sarah Palin. Oops, Kent, the mirror just cracked. But you just proved something I suspected. You're eavesdropping. That's what you're doing all the time you're pretending to do other things, collecting stuff for your diary." Alisia's beauty was in her generous care giving-- and her quick wit.

Then she asked, "What are you doing up so early?"

"I wanted to finish my exercises before I unbolted the doors at 6, just in case Crystal gets here on time to fix my eggs."

"Don't hold your breath," she quipped. "And seeing you all sweaty and shirtless would freak her out."

What does sweet Crystal, our personal chef, look like? She is a pleasingly plump, cherubic young woman who floats around in a fugue state, her face laced with a perpetual pout. Talking to her is like dealing with a somnambulist. She pretends not to understand my Creole and still ignores Jattu, though informed by Steph she was to take guidance from Jattu. I felt like leaving the doors bolted. But I thirsted for our next skirmish in the Crystal wars. Anything to keep my mind off the sadness of facing our clinics. Working in Haiti calls out my gallows humor. But don't be deceived. It's the underlying deadly seriousness of this place that fuels our seemingly innocent by-play. Combat doctors have a peculiar way of keeping their spirits and energy up so they can keep going. MASH, you huskies.

Monday, March 21, Petit Guinee: Chorus Line

When Monday morning rolled around, I found myself doing my clinic rotation for a second week running. I should have been right at home, but home had moved out from under me. Petit Guinee, that beautiful ocean-side clinic overlooking the azure blue Caribbean, had been moved. I had just learned something interesting about the original spot and wanted to check it out. The place used to be an old stripper joint before the earthquake. If I had looked carefully I would have seen risque paintings on the back wall. I couldn't believe I had missed them. Some observant doctor, and some red-blooded American guy I was.

Over the weekend Stephanie had a dispute with the owner of the joint, and had to move the Clinic, giving a special meaning to the concept of 'mobile clinic'. She had secured a lovely new pastoral setting for us, reached from a nearby side street through some banana trees. It was on a grassy spot, right at the edge of some gardens and a vast sugar cane field. The mountains beyond formed a majestic crinkled backdrop. Several cows and goats were tethered in the open, with calves and kids ambling about, sticking close by their mothers. A few stray sows were rooting in the mud between the gardens, and chickens, with their strutting roosters, wandered blithely about. There stood a magnificent gleaming white tent—spacious and open at each end to attract the breeze, with large coconut palms arching above giving intermittent shade. Tessier had been late getting to the Office for transport, and left a message he would meet me at Petit Guinee Clinic. After a few minutes on the road, I realized he might not know of the move and called him. Good thing. He was standing alone in the deserted old clinic building, beginning to wonder what the hell was going on. He was relieved I called to clue him in. I said, "Wait a minute, before you hang up, take a careful look at the back wall, and tell me what you see?"

He began laughing. "Dr. Kent, I never noticed those naked ladies before. How did you know?"

"The grape vine," I said.

He was just down the road, so we arrived at the new site about the same time, walked our equipment through the banana grove, palm fronds acting as bridges over garden ditches. Tessier and I knew from last time we needed to come armed with a folding table and 5 chairs. We set up our teaching consulting room in the back far corner of the tent. Away from the front, where a crowd always gathered, we had a modicum of privacy. Only this morning, given our move and the new location, few people were there at first. After the initial lull, people began filtering in, followed by a stampede. The place was hopping. Our quiet spot in the back proved noisier than we had anticipated. Two vendors, one roasting corn and *patats*, the other selling soda pop under a gaudy umbrella, had opened up stands just around the corner of the tent, but just through the thin tent wall behind us. I was getting used to this now.

As the sun arced overhead and the tent warmed up, I noticed the center of crowd gravity shift from the far end to our end right next to us, each person already grasping their numbered yellow sign-in cards. What was going on? Then I saw it. Our end now had the shadow. They were using it for shade. As a result they had our ringside seats, watching their neighbors talk to a bunch of shrinks. I was just settling in when I heard a loud screeching 'whomp' behind me and felt the ground shake. I jumped. An aftershock finally? No such luck. I heard nervous laughter outside and ran out to see. A huge palm frond was lying on the ground, wedged between the tent and one of the vendors. A part of the frond had raked the tent about where I was sitting. A close call for the vendors (and me). *Off to quite a start,* I thought. *What will come next?* I remembered teaching mental health professionals that paranoia is often quite adaptive, and is in our human defense repertoire for survival purposes. Maybe there was something useful about my increasing paranoia.

Dr. Affricot, my family practitioner for the day, joined us, and we saw our first patient. I had already noticed the attractive teenager, aged 18, waiting outside the tent, because she was wearing a figure-revealing, strikingly stylish red dress, her head adorned by a wide-brimmed white hat. By contrast her face was flat and mask-like, never cracking a smile despite my awkward attempt at a Creole greeting. There was a dreamy quality about her, her movements slow and graceful. I was baffled—and concerned. She seemed so dead and yet so superficially alive. She complained of dizziness, loss of memory, and inattentiveness, but then threw in a hooker. All this began 5 years earlier, when she passed out. This had occurred without warning, no aura, no 'absence', no seizure movements. And her loss of consciousness was prolonged. They carried her from a car into the house, so worried about her that they hastily dropped her on a bed to run call a doctor, not realizing they had overturned a lamp, its hot bulb burning her scalp. She sustained a second-degree burn before they smelled burning hair and noticed. Clearly, her coma must have been deep. Slowly she woke up, had no weakness or paralysis, but didn't recognize people for several days, had retrograde memory problems, and was different from then on, with significant difficulty in short term memory. She now had learning difficulties and inattentiveness. We also learned she was a twin, the other dying at birth.

Looking at her carefully, I noticed her fingernails seemed pale. I suggested checking the inner side of her eyelids. Dr. Affricot had taught me this earlier. His immediate comment, "*chair de poisson*." Tessier translated, "Fish flesh!"—shorthand for the pale skin of severe anemia. On questioning, she revealed she also had severe menometrorrhagia, or bleeding throughout her cycle. Given all the facts, we decided she must have had some sort of hypotensive watershed stroke, or at least some significant insult to her brain, affecting primarily temporal lobe memory areas but also attention. There had not been subsequent syncope or loss of consciousness, and her course had been stable all these years. The earthquake had not made a dent on her, and, as a matter of fact, she didn't even recall it. Though hysteria and fugue states, and twinning phenomena perked through my psychiatric brain, the hard facts spoke to brain problems.

We complimented her on her stunning outfit, which brought only a sliver of a smile, commiserated with her and her mother about her enduring disabilities, and said we could certainly help with her severe anemia, which was news to them. We recommended a visit to an Ob-Gyn doctor for her bleeding, the presumptive cause of her anemia and low platelets, and immediately gave her Iron and multivitamins. Was she a 'sickler', suffering from Sickle Cell Anemia? I flipped madly through my Merck Manual. All I could remember was that in Sickle Cell Anemia, which is inherited, a person's hemoglobin was altered, and caused the red blood corpuscles to change shape so that they didn't fit through the smaller vessels causing pain and damage, even small strokes in the brain. Dr. Affricot already knew all this, and felt not. Mother and daughter were grateful for what we could do. We explained she had had a *'crise vascular'* I or kind of stroke back then, leaving her with the short-term memory problems and inattentiveness they knew only too well. I also mentioned the possible future use of methylphenidate (Ritalin) once she and the world around them stabilized.

I must tell you, Dr. Affricot was good, but my Creole was lousy, not in the speaking of it, but in the understanding of patients, making history taking and translation a laborious, multi-step process akin to slogging through linguistic molasses. I was reminded of the Chinese resident in my 'Dr. Day' residency group, who spoke little English but seemed to come up with good comments at the end, often seeming profound because they were short and incomprehensible. So I tried to relax and drink things in, my personal Zen approach. It worked some of the time but wasn't really in my activist nature. I must have seemed a pest the way I pursued things, but Dr. Affricot was a bright young hotshot, partly trained in Cuba, and he thrived on it, soaking up interview technique, neuropsychiatric diagnosis, and treatment. He was elated, and I was drained—and yet pleased.

I frequently stood up during interviews, leaning on the back of my chair. I have a bad back, and had already lost my ass—pardon the frankness. My 'chair de derriere' was melting away fast, giving me more fanny fatigue each day—this due to substantial fasting, both because of missed lunches and the privation at the hands of Crystal. Probably the sweat lodge was also working its magic. For anyone wanting to lose weight, I wouldn't recommend this overall approach.

Our second case was a 56-year-old woman whose son had died when their house caved in, also burying her store, which was then looted the night after the quake. They had no shelter, food or money at the moment. Her entire world was shaken to its foundations, collapsing around her. Her husband, luckily away at the moment, was spared, but not his van, which was completely crumpled by a falling wall, killing his chauffeur business. Well off and high functioning before, they were now destitute and demoralized. And yet, I sensed a smoldering anger under her sullen depressed veneer, something I filed away. She couldn't sleep, had early morning awakening, and painful recurrent headaches. She found herself crying in gut-wrenching waves, spontaneously calling out her son's name, startling everyone around her. Then she would stop abruptly, and seal over, feeling it undignified. At times she thought she saw him. She had little interest in food, but had not lost much weight. She was isolating herself, hardly talking to family or friends. And she hadn't lifted a constructive finger to help herself or her family.

On closer questioning, we found out they had had no funeral. She finally said they had yet to find his body, something her priest insisted on as the ticket for having a funeral. I pointed out that in the eyes of the Lord, who knew exactly where her dead son was buried, there should be no problem like this. So there was no need to neglect this deeply important consecration, a helpful path to finding necessary closure. I pointed out that other Catholic priests were willing to recognize and observe funerals under her circumstances. Her passivity was striking. I urged Dr. Affricot, who was feeling passive himself in reaction to her awful circumstances, to assign her the homework of beginning to talk to family and friends, and find the right priest.

When he did so, her eyes suddenly flashed with anger and resentment. Dr. Affricot stood his ground, saying she needed to make the effort to honor her son, his death, and the need to find life for the living, as her son would wish her to do. As he said this her anger was suddenly washed away with a flood of tears. Finally, she agreed she would do what he said. Later, I mentioned to him that his passivity mirrored hers and protected them both from her anger, which in turn shielded her from her profound grief. He had found the courage to give her strong medicine and she had responded healthily. He asked her to come back the next week and let us know how things were going.

Dr. Affricot was a little non-pulsed at first, saying, "Your approach was pretty confrontational. I thought she had a major depression."

As we talked, he reflected on her evident underlying anger and realized that was the difference. He did insist we give her just a couple of Diazepam for sleep, which I felt was fine. But I said, "You need to use things like the anger you noticed, perhaps tempered by your own good judgment, to reach some like this." But I agreed with him I wouldn't want to miss a major biological depression, either, and possibly have a suicide on our hands. "You're right. We have to be very careful," I said.

She didn't return, so I had to keep my fingers crossed. Maybe Dr. Affricot would have handled it better with his gentler approach, not losing her to follow up. Or maybe she was just doing better. There was always uncertainty and suspense to our work.

The third case was a woman with classic (overactive autonomic) stress and anxiety disorder. Every noise, every sudden movement, every clack of a table or china, every rumble of a truck in the street, made her jump and scream. She had palpitations and hyperventilation. She even imagined she heard or felt things. Everything seemed like the beginning of the earthquake, which had set off her intense jumpy anxiety, and persisted as if it had happened yesterday. And the subsequent aftershocks, the 'repliques', or repetitions, as the Haitians called them, didn't help at all, just reinforcing her intense startle responses. There is nothing worse than intermittent reinforcement through aftershocks for a condition like this.

She had been in the street when the quake hit, and delayed going home. The suspense as she walked home, destruction all around her, was intolerable and nerve shattering. By the time she got home she was sure everyone was dead, seeing her house completely flattened. But miraculously they had all gotten out unscathed. The reality was clear, and she knew it. But her nervous system, more fragile than some, had already caved in, and desperately needed emotional first aid, and perhaps a tad of anti-anxiety medication to help her right herself and get on the path to emotional recovery. We had to give her several portable self-help techniques that she could use instantly when she sensed any of these things beginning to happen: 'Sac' re-breathing for hyperventilation, the 'Three Breaths', the Valsalva for palpitations, and 'Imagery', 'Progressive Relaxation', and some desensitization 'Sounds and Fears' hierarchy techniques.

By having a patient practice sequential muscle relaxation, building toward total body relaxation, they can be taught to achieve bodily relaxation on demand. When a person becomes anxious over something, they can summon this relaxed body state, making the body state of anxiety diminish or go away. Using this proactively, you can teach an earthquake victim to pair relaxation with increasingly fearful ideas or images in a mental hierarchy, or with actual things that trigger anxiety, such as earthquake-like sounds, trucks making the ground tremble, or even small aftershocks. Practicing reducing these fearful anxious reactions to earthquake related fears and symptoms can interrupt a person's overactive stress and anxiety response, putting them on the road to recover

We stressed that she shouldn't over-diagnose herself, should reality-test her fears, and intervene quickly and firmly when jumpiness set in. She admitted she 'knew' when she was about to get bad', so I said, "Head it off at the very beginning. Get ahead of your fear reaction before it takes you over.   That way you can calm yourself down and give your body and mind a chance to heal." We also gave her just two days of Diazepam, just in case she needed them in a pinch, and told her she ought to hold on to them 'for an emergency' and treat herself with first aid first. " See how long you can get by holding on to them", I said. I said to Dr. Affricot, "Having a Genie in a bottle as a reserve gives more than a '30% placebo effect in my experience. "

I myself have more than a passing interest in these techniques. I have had migraine all my life. A migraine has warning auras, visual and mental. I smell burning rubber, my nasal passages become unusually clear, and I get pleasantly high. These are my early warning signs. Next comes a brilliant expanding visual spot burning a bright expanding star in my visual field, leaving a widening blind spot in its wake. I get scintillating shooting stars, a shimmering descending veil, and even 'gun barrel' vision at times. These are dangerous later warnings about what will be coming next—a blinding headache and severe nausea, which can last hours or days. After suffering through this sequence for many years, starting in my teens, I was fortunate enough to learn progressive relaxation, sac re-breathing, and imagery techniques, practiced in a group setting teaching self-hypnosis for medical conditions. I found I could go to a quiet dark place, or just close my eyes, and relax completely, retreating mentally through imaging to the quietest safest place I knew, and nine times out of ten I could stop the onset of the headache and cause the visual phenomena to go away completely. Only rarely did I have to resort to diazepam.

This success with emotional first aid has given me the kind of conviction and technical skill that allows me to demonstrate these approaches to patients, knowing they are real and effective, and not just phony Band Aids. Conviction and sound techniques pay off for skeptical patients, and practice makes them increasingly effective. But first doctors have to heal themselves of their uncertainty and disbelief, and practice these techniques themselves. As you can here, I was warming up for my next Saturday seminar. Given half a chance, people, and their minds and bodies, could usually right themselves, even in severe catastrophes.

Our second to last patient, a 51-year-old man, on the other hand, had a pre-existing, longer standing, untreated classic major depression. But there was a twist, always a twist, so one had to listen carefully not to miss things. He had mood congruent visual and auditory hallucinations, not accounted for by his periodic overuse of alcohol. There is a real difference between the hallucinations produced by severe depression and those of alcohol, and other brain-based conditions. First of all, 'mood congruent' means that if one is clearly depressed, then the hallucinations (and delusions or unrealistic thoughts that go with them should be depressive, like morbid images or thoughts, and not high, grandiose and elated. That would suggest a manic state. But the hallucinations and delusions usually have a more coherent story line to them, as opposed to organic or brain-based hallucinations, which are often bizarre disconnected kaleidoscopic images. One of the only exceptions is images or hallucinations going with psychomotor epilepsy, which can be more coherent like a TV sequence or a moving picture snippet. This guy's hallucinations fit his down mood.

Our 51-year-old also had a TMJ-like pain (temporo-mandibular joint pain, caused by inflammation of the jaw joint situated behind the ear), giving him headaches that I thought were from anxiety, tension and teeth clenching. That was his complaint. He had stopped working, and was passive and helpless with his wife. He also had some paranoid ideation. His ideas of reference were making him avoid friends and family, causing him to isolate himself. He had no appetite, had lost weight, and suffered early morning awakening. The problematic thing was that he also had more disorientation and confusion than even a more severe depression would account for, confirmed by mental status exam. I was a stickler about doing mental status exams on all these patients, so that the Haitian doctor, and I, could establish cultural norms against which to judge abnormal findings. It paid off in this case. He seemed almost like a Wernicke-Korsakoff, my first thought given his alcoholism, but, like I mentioned before, he didn't babble on with made-up stories, so no confabulation. A brain disorder due to thiamine deficiency, Wernicke's encephalopathy also has confusion, loss of muscle coordination (ataxia), leg tremors, and abnormal eye movements, too. He had none of these. It would have been so nice to just give him the vitamin

Thiamine to solve his problems. We never did know what the cause was, but hoped to find out more after our use of Fluoxetine improved his depression and nutrition, cleared up his hallucinations and delusions, and improved his slowed thinking. Maybe it was due to a mix of all these things, including some other vitamin or mineral deficiency disease, and might be reversible. That would be a godsend. Maybe our medication might help. That's a lot to expect from Fluoxetine, though.

And, we insisted, no alcohol was to be consumed, emphasizing that the medication and alcohol didn't mix, and could be somewhat toxic (I was trying to use a negative placebo effect here). We also pointed out he needed to give his mind and body, and the medicine, the best chance of working. We would reassess him the next week. In this case, the follow-up would be especially important.

Our last patient of the day was a 28-year-old woman with headache, palpitations, flashbacks, a tendency to forget conversations with people, and an urge to run away. She told us she was tormented by jealousy about her boyfriend, acknowledging it was often unfounded, but she couldn't help it. She also had sickle cell anemia. All of this pre-existed the earthquake, and wasn't particularly exacerbated by it. The flashbacks were more like jealous ruminations about her boyfriend, obsessive in nature. Her more long-standing problems, though bothersome to her, were more on a neurotic, or internal conflict, basis. We gave her first aid techniques for her physical symptoms and said she needed to see a psychotherapist when she could, acknowledging it was all very difficult and frustrating.

Coming up for air, I looked around the tent and was treated to a rather amazing scene. Dr. Paul, who had finished early, was filming a real spectacle. Alisia, who had been a Kung Fu instructor in a previous life, had cleared the clinic tables and chairs to one side, and had called a bunch of Haitian kids in front of her. She began demonstrating choreographed Kung Fu fighting moves. The kids were fascinated to see their doctor do this. After demonstrating for a while, she encouraged them, first one and then another, to join opposite her, mimicking her every move. More and more joined in, and pretty soon she had quite a chorus of dancers, all moving in pretty good unison. Parents began filtering in, forming a curious and then a laughing clapping audience. After while, she stopped, chose a few, and showed them how to put their arms and elbows out in front of them, and then begin undulating them, calling this 'The Snake'. Assembling a long line, she got them to do 'the snake' rippling along in quite a long serpentine wave, like in football stadiums, to everyone's delight. Next she had them do 'The Tiger', having them add a loud roar as they advanced on her, and then, when she began to roar, having them turn tail as she advanced on them. They danced back and forth, having a grand old time. I watched all this, thinking this was the best Community Therapy I had seen in a long time.

Looking over at Dr. Paul, I saw something else that touched me. The cutest little girl, Cassandra, age 7, all dolled up in her Sunday best, was standing there, holding his hand. As he explained later, she had been one of the first to arrive that morning with her mother, had sat waiting to see him. When he began and the nurse brought another kid first, she marched right up and demanded her rightful place. Dr. Paul realized his mistake and took her first. Later, he confided in me that she had worms for which he gave her medicine. Almost 75% of the children needed to be dewormed, given their telltale belly pains. When Dr. Paul told Casandra's mother his diagnosis, she recalled seeing a worm crawling out of her nose at night, and recalled she had had them, too, as a girl. Apparently when a child is starving, these creatures get tired of nothing to eat in their empty stomachs, or become overpopulated and can't stomach the competition, so they bail out (the nose or mouth) in hopes of finding food elsewhere. These are the same creatures that all our cute little puppies have in the United States, so you know them well. Sorry for this side trip, but as I got in stride with my psychiatric teaching, I began paying more attention to what my medical colleagues were seeing. Dr. Paul also informed me that as the day wore on, this grateful, infatuated young lady came around to visit him three more times, each time with a cute new outfit on. He had a charming new groupie—though Alisia had him beat with her dance club.

The remaining Haitian staff was hanging around idle, so Dr. Paul pulled out his deck of cards, and began doing card tricks. These were no lightweight amature tricks. As it turned out, Dr. Paul had perfected these and a bag of magic during high school. He made quite a bit of money as a magician during college. He sat down and asked various Haitian staff members to chose a card, any card, and proceeded to do intricate cuts and shuffling maneuvers, every time coming up with the secretly selected card, shown privately to all of us. This was done so smoothly and repeatedly that I heard one Haitian staff member saying that he was no card shark or magician, but a voodoo priest. No one could do card tricks like that without powerful Haitian voodoo magic.

Finally, our Patrol car chauffer came wandering in, watched some of the amazing tricks, and finally insisted we had to go. We traipsed back through the banana trees, and piled into air-conditioned splendor, returning to the Residence. For once, we had finished early, maybe because patients had not heard of our new location, and we were famished. Ever hopeful, we looked forward to Crystal's late lunch. Unfortunately, except for some dried out rice and beans, and a few plantain fritters, it had been devoured by someone. Our next hope lay in the dinner menu. Dr. Paul said his wife, a doctor also, had come to Haiti and had worked here in Petit Goave a month before him, had told him he should ask for goat meat while here because it was delicious. We started telling Crystal our desires at every chance. Stephanie had no interest in such things, being a vegetarian, but we all did. We came out to the kitchen and discovered she was again cooking chicken in red palm oil sauce.

I was too embarrassed to mention this until now, but during the first few days in Port-au-Prince I thought I was catching a cold. Then my left Eustachian tube, (that connecting tube between your middle ear and your throat), opened up suddenly, causing my eardrum to pop with every breath. Before this, it only happened on airplanes and on very hot days during strenuous tennis. 'Hot days' are the operative words here. Sweating during my first mobile clinic tent sweat lodges, the heat had gotten to my ear. I soon discovered why tropical men leave their shirts out. Air conditioning! Then something more ominous happened. I began to notice this awful smell in my nose and felt a nasal infection creeping through my nostrils. *My god*, I thought, *I'm beginning to have a serious sinus infection. This could get out of hand.* But I wasn't sure and didn't want to overreact or sound the alarm to the medical types all around me.

Doctors are the worst patients, either over-diagnosing or denying their illness. Mind you, doctors are even worse patients when they are treating themselves, and maybe especially when they are shrinks. They over diagnose, then under diagnose, first fearing it is in their bodies, then worrying it is all in their heads, then getting stuck in an obsessive quandary somewhere in between. As a result they end up embarrassed about taking too long before they finally tell a medical doctor.

Luckily, given my nasal condition, I had some nasal spray that I had stealthily put some ear antibiotic drops into, just in case something cropped up. Luckily, a spritz or two of my nasal concoction cleared up my smelly nose. I guess I did have something going on there, not just a figment of my imagination. What I didn't understand then was the stench of death surrounding us all was feeding my paranoia. I wasn't yet thinking, just experiencing at that early stage in Haiti. I was floating on the thin veneer of my rational mind somewhere between dream and nightmare.

In the same vein, toward the end of that first week in Port-au-Prince, I felt an aching in my armpit. I noticed a tender boil in my left pit. I was surprised to find myself getting quite anxious, remembering how a little elbow infection had gotten rapidly out of control in Paris, ballooning into a seriously infected elbow requiring hospitalization, surgical drainage, and major antibiotics. What would happen here in Haiti if my armpit blew up into a massive infection? They couldn't just cut my armpit out. Worse, where would they take me? I felt dizzy. Then I remembered I had been told to bring tubes of triple antibiotic and antifungal cream. Whipping the antibiotic out, I put a little dab right on the miniscule pimple and breathed a sigh of relief—for an hour or two. I remembered my dad humming the Brill Cream commercial, "A little dab'l do ya". But I began watching it like a hawk. Thank god the next day I awoke feeling comfortable. My self-pity pain was gone. I took a peak. Yesiree, it was really gone, just a tiny red dot marking the spot. I was home free; not realizing it was the second telltale sign of another insidious disease.

Later, here in Petit Goave, after about a week of mobile clinic sweat-lodging, I slowly became aware of something else, right after seeing that girl with festering scabies on her left thigh. It really disturbed me—a hot itching burning feeling on my inner thigh right up near you know what. As the day went on, it became worse, and so did my concern. Finally that evening I told Dr. Paul about it. "I don't need to take a look," he said quickly, (or maybe he didn't want to). A brilliant pediatric hematologist-oncologist, he was now completely boned up on his tropical medicine. I awaited his pronouncement with apprehension. There I stood, telling him I had an infection down *there*. He looked at me, paused dramatically, and said, "Kent I think you have a very serious condition. It's called 'Crotch Rot', from all the sweaty sitting you're doing. In more genteel circles, they would call it 'Jock Itch'. Just hit it with some anti-fungal cream, and you'll do just fine." You'll be relieved to know he was absolutely right—but at the time I felt suddenly exposed. He had seen my unconscious again. But after first blush, I finally began to think, *What the hell is going on with me? And, really, with all of us?*

Something else caught my eye. The volunteer doctors were all walking around with their newest medical prowess dangling from their belts, about crotch level—upside down plastic bottles of Purell, hanging there like some defrocked codpiece next to their large pendant collection of medical instruments. THEY were prepared for anything. THEIR Purell, the newest mass Haiti donation from the American pharmaceutical industry (casting purifying pearls before Swine Flu, craftily catering to good medical practice and medical paranoia) hung there conspicuously, for all the world looking like elongated plastic testicles. Not to be outdone, soon the Haitian doctors had them too. And even the female physicians displayed theirs proudly. They all suffered from a shared mania, a hand-washing fetish. Not that I don't believe in cleanliness between patients, for all our sakes. But I noticed many of them forgot to do that occasionally. But they never forgot their pendant Purell. Especially Dr. Paul. He was fastidious.

But where was my psychiatric version of Purell, and why didn't the Pharmaceutical Industry cater to me? What was I? Chopped liver? They should give me my own Purell, emblazoned "Pure-all for the Unconscious". Then we would all be safe.

But truth be told, what a fine lot we ALL were. Scratch any one of us doctors and you get fears of fulminate infection, scratch a psychiatrist and you get fears of rampant paranoia. You couldn't lie down in Haiti or go barefoot lest you get nematodes or hookworms, couldn't scratch yourself after touching patients for fear of scabies, nor put your dirty fingers in your hair because of Tinea Capitis, or to your lips and face because of Thrush or Impetigo, or touch that tear in your eye lest you get bacterial conjunctivitis; breath the breath of your patients and not get TB; touch their blood and not get HIV; nor let a mosquito bite you lest Dengue or Malaria get spit into your blood stream. And that dangerous drinking water might give you dysentery, the incompletely cooked pork contributing

Trichinosis. And then if you dared to sleep, per chance to dream, you might have nightmares or worse, primitive fantasies and delusions, maybe even auditory hallucinations of roosters and dogs running rampant through your mind's ear, or spiders running across your bare feet. And each thing that goes bump in the night might be an aftershock. I knew so little when I was last in Haiti that I was naively and fearlessly brave. Back then I was the sweet bird of youth, flying high above it all. But now, as a grey haired bird long of beak, I knew too much, or thought I did. One thing I was glad of--I had gone to Harvard Medical School back when they still taught you how to listen to patients and pay attention to their bodies, not immediately sending people off for fancy tests and procedures. MRI's are hard to come by in the Haitian boonies.

None of us, underneath, were resting easy, because we all shared the sensation of floating precariously on the seething magma of our desperate patients each weighing on our fragile minds, with uncertain terra firma under foot--tectonic plates sliding inexorably toward some physical or mental disaster—that is, if we didn't use Purell or Clorox, and especially the DSM-IV. You could hold that psychiatric bible up and exorcise any mental demons lurking nearby. What I hadn't counted on was how much my old Merck Manual would also become my bible once again. It felt like I was taking Part II of the medical boards, neurology, and tropical medicine, all over again. I had to clean out my cobwebs quickly, because the heart of darkness was never far away.

We had all caught the terror of the Haitian earthquake, with her aftershocks and rampant dis-ease. Unwittingly, we had become our patients. And through this process, this human contagion, we were taking the measure of their plight, giving us humble appreciation and empathy for all they were suffering. Doctor, heal thyself, and then you might have a chance to heal your patients. Nobody can escape Richter Fever and the impoverishing vermin of human tragedy. Earthquakes level us all.

I came to realize that rationality was just that thin veneer I mentioned, covering the surface of our minds in Haiti, as we walked among our patients, shuffling along the surface of our mental world, our denial barely working to protect us from the rampant infection all around, both mental and microbial, as we trusted the fragile earth below us and her thin earthen surface, only momentarily stable-- the risk of primitive earthquakes not far below our own medical crusts. We talk blithely of their inadequate infrastructure. What about our own?

I capped off Monday, the 22nd, with a brief return to Notre Dame hospital, looking for Dr. Lynda, to see if there where any takers for my proffered hospital resident seminar. She still hadn't had time to deal with it, was late in doing the on-call schedule, and politely blew me off. I wound my way around trying to find some of the residents myself, thinking I could recognize them since two had been in my seminar, along with some nurses. In doing so, I met Brita again, that bright cheery Scandinavian Red Cross worker serving the tent cities. She listened to my proselytizing spiel a little more attentively than others, though still seemed to be at a loss as to what I might be good for. My final stop was the hospital Red Cross OPD tent, where I was pleased to discover two of the hospital nurses I had taught. They waved hello to me and said they would deliver a note about my offer to teach. I said, in parting, "Don't forget to come this Saturday!" In unison, they said, "We'll be there."

I called Samedi to be picked up at the Hospital gate, plotting to get my coca-cola fix. Casing the street for would-be abductors, I stepped outside the gate and was immediately accosted. I had been through this before, outside Mars and Kline Mental Hospital in Port-au-Prince, but was unprepared for it in Petit Goave. Haitians hawking 'primitive paintings' descended on me. Figuring doctor-types, especially a 'blanc', to have dough, they pounced on me as fair game. They were surprised when I asked in Creole how they were doing, complimenting them on their wares. Then I asked them if they could lend me some change for a Haitian Cola, since I had forgotten my wallet. That made them melt away quickly. But not my reverie.

When I was first in Haiti, my mentor, Odette Mennesson-Rigaud, introduced me to painter Andre Pierre. I didn't have enough money back then to buy any of his fascinating works. His paintings would bring a mint now. I did talk him into painting some Govi jars, and a set of 3 Rada drums that our mutual friend, Coyotte, the drummer, had made for me. Andre adorned them with beautiful *veve* drawings, the cabalistic designs through which the gods are summoned.

Only the driver's honking brought me back to my senses. He was surprised to hear my daydream and learn I helped sponsor one of the Haiti's early U.S. primitive art exhibitions, back at Seton Hall University in the '60s.

I had skipped lunch already, drank my cool coke, and was happy to get back to the Residence. Why wasn't I more hungry? My stomach was really shrinking, not a bad thing, putting me somehow past the point of hunger, which had peaked hours before. I examined the fly-protecting plastic latticed covers, placed over all the food, prepared hours before by Crystal, who left well before 5. I refrained from pulling them all off dramatically, like some high-class waiter. I didn't want to disturb all the flies. We were all arriving late, about the same time. Suddenly my hunger reared its teeth and I joined everyone as we descended on the food. When the covers were whisked off, there was a familiar collective groan. Tired of chicken, we had been asking for grilled goat for some time now. But there it was again, a few scraggly pieces of overcooked chicken. Originally we had liked the tasty tomato flavored palm oil sauce, and the slightly off-tasting red beans and rice. Now the bloom was off the rose. We dug in out of habit, and desperation, but the meal wasn't sufficient, or satisfying. Only the thought of all those children starving in China (and Haiti) held my tongue. In the mornings, Tom and I were driven to cooking our own eggs.

Tuesday, March 23, Miragoane: Rendezvous with the Past

Today I went to Miragoane again, and was not looking forward to that long rutted road. To my surprise, probably because of daydreams, we arrived at the Mirogane Clinic in no time. We were armed with chairs this time. Out came my inflatable pillow, and we were in business. Our first patient, 29, talked of his profound sadness and recurrent thoughts of killing himself. As he spoke, he seemed to be swallowing his tongue. He told of being unable to work for the last 8 years, but still hoped to pursue his dream of running a business. His speech was so garbled we had to stop him, asking him to open his mouth. His tongue was chewed into a large bulbous cauliflower. "How did that happen?" we asked. "Every time I have an attack, my neck and head go into spasm, my jaws clench clamping down on my tongue, and I can't stop chewing on it--maybe two or three times a week. The only thing that stops it is 5 pills of 'Akineton'. *But what the hell is Akineton?* I thought, *and what could this be? He doesn't fit Lesch-Neyhans Syndrome* that self-mutilating lip and tongue chewing developmental condition. That starts at birth and the person in severely mentally delayed. Dr. George and I were stumped. *It sounds like a dystonic movement disorder, or maybe a seizure?* We pressed on. Every case was a cliffhanger, one way or another, and we needed more information. I always felt a little anxious as I hear the beginning of a person's story, wondering if I could figure it out; whether I'd come up with something. Every case is compelling in some way, and I wanted to help if I could.

His life had gone well until he was 20, but he had never worked after that. "What happened back then?" we wondered.

"I became violent, got in fights, had to leave school."

"Then what?"

"I was hospitalized."

"Were you thinking everyone was against you, maybe even hearing voices?" "How did you know? Yes, it's painful to remember. But thank the lord it never happened again. But ever since then I've had these attacks and am too depressed to work."

Suddenly we had our diagnosis, sad to say. I whispered, "Tardive Dyskinesia" to Dr. George and his eyebrows went up. "It's that dreadful permanent side effect of the neuroleptic anti-psychotic drugs they gave him to cure his paranoid psychotic break."

Dr. George said, "Don't ever use anti-psychotic drugs again, and look, this kind of condition tends to go on, but we can help you manage it better. We think you ought to try taking an anti-dystonia (anti-Parkinson) drug regularly, and see if we can cut the attacks way down, which would help your situational depression and give you a chance for your dream. Here, try this Kemadrin daily, and come back next week, tell us how you're doing. You can hold on to your Akineton in case you have a full attack."

He was slightly encouraged for the moment, and we were guardedly hopeful. This was a tough situation. If this didn't work, we would try Carbamazepine, in the off chance an anti-seizure med would help these dystonic, possibly myoclonic attacks. Repetitive spasmodic muscular spasms, or jerks, are also helped by Carbamazepine. I knew in the back of my mind that other drugs could target various symptoms and conditions better that we were seeing IF we had them, but at least we could do something in most cases. I was also noticing we were attracting more chronic patients now, while seeing fewer acute quake-related cases. Our patient population was shifting perceptibly. What to make of that? I would have to step back and think about this pattern when I could catch my breath.

The next case, a 25-year-old woman, fit in with this trend. Trying to pursue nursing, she had had three psychotic breaks, each at stressful career junctures, treated with Haldol, and later, Risperidone (a newer generation of anti-psychotic drug). Now at this point, she seemed flattened and surprisingly confused. She had a strange echolalia, repeating our questions 8 or 9 times. She wasn't quite catatonic, and kept saying, "They're coming to get me." She had lost her meds, and psychiatrist, in the earthquake, and was deteriorating. Because there was a history of highs too, we hoped her peculiar state represented a less serious psychotic paranoid depressive phase of a manic-depressive illness, now re-emerging because of the earthquake, and loss of meds and psychiatrist, rather than some permanent neurological condition. Because it had a mood disorder aspect, we decided to try her on Carbamazepine, especially since she refused any Haldol because previous massive weight gain caused by that drug in the past. If you are wondering why you keep hearing about the same medications over and over again, and seemingly for such different conditions, it is because we only had available one or two drugs in each category, and some--like Carbamazepine (Tegritol)--are good for several conditions: bipolar, psychotic and seizure disorders. We're realists, not just Johnny on-notes, trying to fit all our patients to a Procrustian couch.

After we saw a 51-year-old woman with a classic depression with some added physical symptoms, treated with Amitriptylene, we met a bright-eyed manic woman with a paranoid tinge coupled with fleeting visual and auditory hallucinations who jabbered at us a mile-a-minute. We gave her, as you might guess, Carbamazepine. What really helped, though, was our follow-up clinic, where we could see any improvement, or lack thereof, and adjust or change meds as needed. This lady, though she was still a little high the next week and wanted our names so she could thank God in church for us, was in fact cooling off, with no voices and visions, and now eating and sleeping better. Her confusion was clearing also. We were thankful. We upped her Carbamazepine a little. Close follow-up is instructive, corrective and reassuring for us all.

Our last patient was 14, still in first grade, constantly embarrassed, and pestered by his mother and teacher for not concentrating and working hard enough. Miraculously, he was not yet a troublemaker. A handsome lad, he was miserable, and yet a little cocky still. His older brother had the same problem, and finally, when she admitted it, so did his mother. But she said she had to endure school so why shouldn't he? We talked heredity and common sense, suggesting he was generally a smart proud boy, his self-esteem and morale not too damaged yet, with clear circumscribed troubles only with reading, writing, and arithmetic. So, instead of wasting his time and ruining his confidence, he (and they) should make a bold, pre-emptive move and seek out some good interesting apprenticeship, find his calling, and get a jump on his life's career. A lot of people were out of work and out of school now, so making a change like that wouldn't attract attention. And because he had street smarts, good looks and style, he could do well, maybe have an edge over other kids. A smile crept over his face, and a light bulb went off in mother's head. We said goodbye and wished them well.

Coming up for water and a stretch, I looked around. Where was he, our jovial alcoholic? I had really hoped he would meet our challenge and come back to see us, ready and competent for a more thorough evaluation. But he was nowhere to be seen. I was disappointed but not surprised. I wasn't very effective with alcoholics. Only a few people I know really have the knack, and maybe not in times like these. I tried not to be too hard on myself.

Today was a lighter load and thankfully went quicker. I had a very important date to keep. My return driver was already approaching when I called Samedi. He and I had had a heart-to-heart early that morning, and he knew how important my rendezvous at Alice's IMC Leogane headquarters was to me. He knew of the friends I had lost there in Brache and Masson, my old field site, and how eager I was to meet Pierre d'Haiti, Karen Richman's foster son, and Charlie Fangala, her godson. I had so despaired of making any contact that this breakthrough opportunity made my heart sing. What was happening to me? I had gone from vowing brashly to get to Masson no matter what to settling for a near miss.

I was so lost in fantasy I was surprised when we got to Petit Goave in record time. The mental space on drives had become another way to decompress and reflect. I realized I'd be too early for Leogane, so we cruised by Stephanie's office, and stopped briefly at the Residence. Crystal was shocked to see me so early and out of the blue. And even better, I found uneaten lunch sitting on the table. I couldn't figure out who it was for, sitting there warm and enticing, but possession is nine/tenths of the law and I dug in. It was the same old chicken and rice with read beans. But it was fresh and I was famished. There was an air of mystery about Crystal's tactics and timing. Anyway, this was turning out to be a good day. I wolfed it down in 15 minutes, grabbed a few things, and hopped in the waiting Patrol car. Just as we were leaving another Patrol car was pulling in with no passengers. I wondered why? But we had an important mission now and hotfooted it on to Leogane.

On the way I was blind-sided, first by Alice, that lovely new French director at Leogane, "Are you still coming today?"

"Absolutely! I'm on my way."

"Oh, okay, but they've called me to a Leogane NGO meeting so I won't be there."

"You won't?"

"The guard knows you're coming."

Merde, I thought. *Now I can't help my friends.*

"I may make it back before you leave, though. I hope so."

"Me, too."

As we approached Leogane, it all came back, erupting back into my head and heart when I saw all the destruction. On the way to Petit Goave the first time we had driven through the outskirts of Leogane, and I had been so upset passing Brache itself, that the massive scope of Leogane's destruction had not fully registered. I also suspect that if I had taken it in it was so painful my mind deep-six'ed it. But now it flooded me. Because we were lost in back streets at one point, and went deeper into the heart of Leogane, I couldn't avoid the awful truth. I had been told that Leogane, the epicenter of the earthquake and afterquakes, had been 90% destroyed. And it was true. Everywhere I looked, I saw collapsed buildings. The driver had never been to the new office, since the building had just been leased and set up. We wandered around town looking, exposing us to far more than I had expected. Though some spotty clean up had begun and stray building survived, the place was a wreck, and tents and leant-to's of every description were crammed in everywhere.

Though half the population had died or fled, the city was teeming with people, women doing commerce, their brightly colored stands doting the roadside between piles of refuse and broken walls, with goats, pigs and chickens hunting and pecking for anything they could find. Dogs scavenged everywhere, dodging brightly colored camions piled high with sacs of produce and charcoal, with people perched precariously on top.

Life was everywhere, and death was buried below. For all the world it was like a city gone crazy, more like a New Orleans funeral than a dirge. I was wracked with feelings, ranging from sadness to admiration. How could they do it? But how could they not? Life had to go on, even in the midst of death and destruction. I had become so used to things in Petit Goave that the freshness of all this hit me like a ton of bricks.

The only thing that buoyed my feelings was the prospect of seeing Pierre and Charlie—if we could only find the place. It took forever but we were only a little late, and I worried they might be waiting. I had told them the address during my original calls, but now that I saw how hard it was to find, I was worried they'd never get there. What if I came all the way to Leogane and never saw them? This plan was already breaking my heart, my settling for a little and getting nothing. My driver rolled down his window, hailing a local guy on the street, "Hey, bossman, where's the new NGO IMC Office. We know it's nearby?"

"Just take a right down there beyond that huge sign."

I squinted out the windshield into the glare of the afternoon sun, and saw a fat 30-foot pole going out of sight up into the air. Peering out the side window, I could see it was holding up a huge empty billboard. After turning and going halfway down the side street, the driver began honking, the usual drill to Open Sesame. On cue, this big red wrought Iron gate slides open and a uniformed guard with cool shades and an ugly sawed-off shotgun stepped out, squaring off to face us. Seeing the white Nisan patrol car bearing down on him, he lowered his gun and waved us in. Thank god Alice had tipped him off.

We arrived and I hopped out. The place was deserted. I asked the guard to call my friends with the precise directions. He began dialing away on my cell phone, apparently having some trouble. Meanwhile, I had something very important to do. I knew Alice was French, and not a vegetarian, so would have her cupboards well stocked. I ran back, almost salivating. I was right. Out came the peanut butter and jelly, and some crackers to indulge my bulimia. I hadn't realized how pent up I was. Then I pilfered her cabinets for all the Power Bars, peanut butter crackers, and cookie packs I dared take, stuffing them into my handy fanny pack to the point of bursting. I had trouble zipping it. I knew she had four new medical volunteers to feed, but I had four, myself, back at our own Residence. Paul, Alisia, Melissa and Dr. Alice would be ecstatic—if Tom and Jattu didn't have at them first. I had already done my damage.

Just then I heard the volunteers arriving, surprised to see only two. And people I knew, my previous Barbancout buddies from Petit Goave. "Where are Kathleen and Laurie?"

"We had a terrible gas explosion here last night, and Laurie caught it full blast in the face. She's gone to Port-au-Prince to be evacuated. Kathleen went with her."

"How bad off was she?"

"Second-degree burns on her face. Hair and eyebrows singed off; she was blown back against us. Some guy installed it wrong. She was traumatized. We all were. But I think she'll be okay."

Stephanie's constant worry about gas made more sense. I should have known, since we had some trouble with propane in Fanghetto, our Italian vacation place. Maybe that's why the Italians call their gas cylinders 'bombolas'.

After chatting with them, I heard Pierre and Charlie arriving. Pierre was older, a dignified young man with a great smile, and smooth charisma. He had started a school nearby, and even built a residence for disadvantaged Leogane children needing education and lodging. The residence had been damaged in the earthquake. He had started a fund-raising tax-exempt foundation for his work, and had been able to pick up a nice property with a warehouse-like building on it. He was hoping to rebuild the residence. Charlie, a bright attractive younger guy had lost his job in the quake, and was looking for work. To my surprise, Pierre had a very successful high-end carpet installation business back in Virginia. He had even installed carpeting for the Red Cross headquarters, getting to know the president, who gave him a nice donation.

"I'm sure I can't match that but I have something I brought from Paris for you and Masson." I pulled out my billfold, and gave him $240 US dollars. "My wife and I wanted you to have all the cash we had."

"Thanks. We'll put it to good use."

I found myself wishing it were $1000. Then I turned to both of them, "I don't know what will come of this, but I hope Alice comes in while you're still here. She may be hiring for various things. She needs to know people she can trust who know the Leogane area."

At this point I brought out my MacBook for a little surprise. showing them pictures of Masson some 50 years ago. Little Tonio and Elminar, kids of Joselia and Ternvil Calixte, the Voodoo priest I stayed with, would now be in their fifties--their parents and grandparents long gone. Thinking this, my eyes misted, not just for them but for the countless people now dead because of the earthquake. They saw me get quiet, with a tear running down my cheek. We shared a moment of silence together, then talked of the damage in Brache and Masson. They said it was just as well I didn't come see because it would bring me even closer to so much sad loss.

We all looked up when Alice arrived, accompanied by another familiar face from Port-au-Prince, Mbassi, now second in command for the three new mobile Clinic start-ups in Leogane. Gressier had been placed in their bailiwick. Pierre and

Charlie were glad to hear about medical care out-reach finally coming to their area. And Alice was glad to meet them, promising to see what she could do for them down the line. Though less immediate a homecoming than I had hoped for, I felt my most important private mission in Haiti had been accomplished, and with Stephanie's and Alice's blessing. The only big remaining question mark was my little four-year-old, Makenta, in that preschool orphanage outside of Port-au-Prince, a question I feared might go unanswered. But I was grateful for what the day had brought me.

Then I reached back to my fanny pack to get my camera for pictures.

No camera. I guess thievery doesn't pay. I must not have zipped my fanny pack sufficiently after stuffing all the Power Bars into it. I quickly retraced my steps to the kitchen. It had fell out, landing on the terracotta floor. I picked it up and saw a weird iridescent discoloration in the corner of the screen. But worse, the lens mechanism didn't work.

With a sinking feeling, I realized I wouldn't be able to take pictures of them, or anything else on the rest of the trip. I was beginning to feel jinxed—first my glasses and now my trusty camera. But one thing I did have was a feeling of gratitude toward Stephanie. Originally I had been quite down on her around not visiting my field site at some point. But all things are relative, and as I came to fully inhabit the risky new world we were living in, I found myself deeply moved by her finding a safe, realistic way for me to see my Leogane friends, and visit the rest at least indirectly. Pierre and Charlie promised to carry word back to those still alive.

Wednesday, March 24, Platon: Unusual Medicine

I woke up early Wednesday glad we were going back to Platon, looking forward to the boat ride. Dawn revealed a clear sky, assuring me I would be able to carry out my mission. I had felt guilty about having missed the boat with our mad woman of Platon the first round, her words still ringing in my ears, "Give me some food, build me a house"—all fundamental tenets of IMC care. I had asked dear Crystal if she would buy me a big bag of rice, another of red beans, and a nice bottle of oil. For once she did my bidding, though coming back with less than I had expected. I also procured a big bottle of multivitamins. But this wasn't the most important thing. I had that tent that Norbert had generously donated for some needful Haitian. And I knew just who that would be. I made room in the tent bag, putting in the food and vitamins, slipping in a poncho for good measure.

I was happy that Drs. Paul and Alice were coming with us that day, and even more so when a little 4-year-old showed up boat-side with a bad burn, now somewhat infected. I mentioned to Dr. Paul that people often showed up at the time of our launch or return, so we often ran brief impromptu clinics before and after. So it was all right to help this hopeful mother and daughter. With little prompting, these two fine docs moved right in to clean up and dress her injury, putting emollient and anti-biotic cream on her, and, with great difficulty, getting her to swallow an oral anti-biotic. She cried and cried. Then, when it was all over, she looked up at Dr. Paul and gave him a big smile. She knew her torturer was helping her all the time (or at least that he had stopped). I borrowed Alice's camera and took some pictures. She lent it to me periodically after that.

We were motoring across the Petit Goave bay, the mountains lying off to the left like the spine of a sleeping dragon.

I saw something big leave the Spanish hospital ship to my right, reminding me of what I had seen coming back from Petit Guinee clinic with our driver. We had run into a huge Spanish armada lumbering toward us on a narrow dirt street. We swerved hard against a wall to make way for them. First came a wide squat camo hummer, followed by a monstrous massively tired dump truck, pulling a trailer hauling a Stegosaurus of a bulldozer. Running caboose was a big amphibious troop carrier filled with camo soldiers, all strangely wearing hard hats. "What the hell is that," I asked the driver, "Where are they going and how did they get here?"

"They're cleaning up the remains of a collapsed Haitian hospital," he said, "and building a new one for us. They're serious folk, doing a great job. But I don't know how they get here."

On the boat that morning, I watched what turned out to be a huge Spanish landing craft moving across behind us heading for Petit Guinee. *So that's how they do it, get all that big earthmoving equipment to Petit Guinee,* I thought to myself. We saw lots of NGO cavalcades cruising around, all seeming so pompous and self-important, as they went roaring along, honking conspicuous, like our big Nissans. Once again I thought how we ourselves must look. But this Spanish Armada won the phallic NGO derby, hands down. What got them off the hook was the good work they were doing. Maybe us, too.

The Spanish weren't the only invaders today. The Red Cross invaded my boat ride too that morning, their frantic cell phone call piercing my reverie. It took me a moment to believe it was actually ringing in my pocket way out at sea. "Dr. Kent, Dr. Kent, we have this psycho patient, really delirious and crazy. He's 20. Can you come see him for us?" It seems Brita and the Red Cross had finally figured out what I might be good for.

"Sure, uh, Well, I'm out in the middle of the bay on the way to our boat clinic in Platon, but if you can figure a safe way to hold the patient, I'll come over immediately at 4 o'clock. Be happy to see him." This was starting to get interesting.

Not twenty minutes later, as we were passing Bananier, I got a second frantic call, again from Brita and the Red Cross. "Sorry to bother you again, Kent, but we've got a second patient. They seem to be coming out of the woodwork. This one's a 61-year-old woman who is incredible agitated, muttering about doom and destruction. Her family can't handle her any longer. Can you see her too?"

"Have them bring her to you at 5. Sure, I'll see her. Do you have a place for me?"

"Absolutely, we have this nice little tent, all ready for you." So now instead of refusing to let us pitch our tent on their hospital space, they were offering us *their* tent. Things were really looking up; it would be a very late lunch today. I filled Tessier in on our new Notre Dame Clinic and warned we'd be having an unexpectedly extended afternoon.

When we arrived at Platon, I had to jump in the water again, rolling my pant legs high to show the assembled villagers my support hose. But I kept my precious medical contraband high and dry, passing it on to someone wading out to help us. The usual crowd was waiting, delighted to see three leaping white doctors instead of just one. Tessier and I reclaimed our Tamarind tree, loving its high branches and cool shade, its leaves casting mottled ever-shifting shadows over us. The day before I had put out the word through the boat nurses to make sure that Alice St. Leger, our madwoman of Platon, would keep her follow-up appointment. And there she was, circling our spot bright and early, minus her ax-hoe, a good sign that rapport was improving. She marched right over once she saw our chairs up, sticking out her fist to do an Obama knuckle greeting. We assured her we were doing follow-ups first, and that she would be number one.

But first we had to get our new Haitian physician, Dr. Philogen, seated and oriented, letting him know that brief follow-ups would be done by his compatriot. Dr. Beauge had seen a few cases the week before, so we planned continuity of care and learning. Dr. Beauge was out front orchestrating beginning patient care, while Drs. Alice and Paul set up shop to my far left and right, *en plein air* also. When Dr. Beauge heard his patient was back, he popped over, and we filled Dr. Philogen in. Alice St. Leger was so eager she could hardly sit down, but still had a blood pressure of 200/120. We had been too gentle, and so hit her with a more powerful diuretic, Furosamide. When it first came on the market it sounded like a dangerous angry drug, wringing people's bodily fluids out too vigorously. But now we had tamed it. We knew how to use it. If we could get her to pee off some of her fluid volume and salt, the blood fluid load on her heart and vessels would be less and her blood pressure would drop more. But easy does it. We didn't want her to stroke out from low blood pressure.

Then I asked her, "Alice, you remember asking me to build you a house and give you food. And I didn't listen too well." "Sure do," she said. 'Well, in this bag I have both things you asked for, plus some vitamins." She thought I was kidding until she saw the tent, and the red beans and rice. She began jumping up and down, and a crowd gathered to see what was going on. I was worried about envy, but they all seemed genuinely surprised and pleased for her. I beckoned her relative. He was a big strapping middle-aged man. I asked if he would help her put the tent up, and make sure it didn't walk off somehow. He promised he would. After arranging all this, we asked her to return next week for follow-up. We had to make sure her pressure came into healthy range. Next we saw our depressed mother with twins who was clearly much improved, looking much better than on the front cover if this book. Finally, we saw a seizure follow-up, slightly improved, needing increased medication.

Our first new patient was a rather irritable impatient 60-year-old gentleman, dressed in a tattered sport coat. Grumpily settling down, he told us his wife died two months before the earthquake, which destroyed the new house he had been building for her. They had already lost 3 of their 6 children some two decades ago. Before she died, they both cried regularly together about their losses. He had only a few soft depressive signs, but was isolating himself, driving people away with his irritability. We pointed out he had a healthy capacity for grieving but had sadly lost his grieving partner, and needed to overcome his grouchy standoffishness and find a family member, also grieving, to share his regular waves of sadness. We also gave him something to take the edge off his anxious anger, improve his sleep, and facilitate social interaction. It was Dr Philogene's 'secret potent No. 9', called Phospate of Codeine. I cautioned about its addictive potential, but since we had no diazepam, it sufficed, a few drops a day for just a week until he came back. He hurrumphed, said a grudging goodbye, and shuffled off.

A 24-year-old with clear warning aura, dizziness, loss of awareness, and 'falling down' once or twice a week required our seizure medication, Carbamazepine. Then in came another 24-year-old seizure patient, normal until age 6, but now mentally retarded because of probable H. Influenza meningitis. She was elfin, confused and chattering all the time, a chronic problem. But her big problem was agitation and running away, mumbling to herself all the way down the path, often followed by a hooting crowd. We zeroed in more, and finally clarified she had already been developing slowly, and had had occasional seizures, though none recently. We wondered about subclinical ongoing seizure activity, and felt we should cover all bets by using Phenobarbital, good for seizures, agitation, impulsivity, and motor mouth tendencies, starting at a low dose. Follow-up showed some improvement, so we upped it a little the next week.

Our last patient was a 26-year-old woman with severe episodic headaches, and even more bothersome, total-body tension and muscle aches. She had moments like this before the earthquake, but after her house caved in, killing her mother, and injuring a child, she became acutely worse. She had slightly high blood pressure the first time we took it ('white coat', or doctor visit induced high blood pressure?). But something about her bouts of intense bodily tension and headaches made me wonder about labile hypertension, episodic bursts of high blood pressure, masked by periods of normal pressure. Maybe her headaches reflected more than muscular tension in her scalp muscles. Maybe she had periodic high blood pressure headaches? We taught her relaxation and imagery techniques, but the next week the bouts of severe tension persisted, and we caught a moment of higher pressure. We felt a little Atenolol was in order to take the edge off her extreme muscular tension. Beta-Blockers like Atenolol help reduce muscular tension, reducing tension especially in smooth muscles. Because of the smooth muscle relaxation, they reduce and stabilize blood pressure nicely. The walls of blood vessels are made up of smooth muscle. By the following week she was better and talking to people about family losses, and able to sit still long enough to do some emotional work. Tears were finally starting to flow.

Just as we were finishing, a strange thing happened—only in a fishing village. This guy came running up, apparently wanting his share of our attention. He insisted on showing me his prize catch of the day. Always mindful of good town-gown relations, I obliged.

The wind was coming up and dark clouds were gathering on the horizon, *maybe over Notre Dame hospital as well*, I thought. So we had to cut things a little short to pack up and ship out. Being a fishing village, when they felt the wind switch around and saw black clouds scudding across the sky, they knew why we were weary and had to leave in a hurry. Down at the boat, Dr. Beauge was giving some kids a lobster biology lesson.

As we stepped into the boat to go back to Petit Goave, one of the boat clinic nurses pulled me aside, whispering, "I was so pleased what you did for your patient. We all were. But what about me?"

"I haven't forgotten you, Eustache," I said. "I have something in mind for you. I hope I can pull it off." I knew she had no shelter, not even a tent. She smiled shyly, and began chatting with Marie, the other nurse.

Once under way, I realized I had made a big mistake. I had gotten into the slow boat, just to be friendly with Eustache. I watched the other boat pull steadily away ahead of us. I began looking at my watch. It was fast approaching a point where 4 o'clock wouldn't happen, and I didn't want to be late for my Red Cross demand performance. The other boat kept pulling farther and farther ahead, and finally I had to resort to my Zen defense, trying to practice what I preached, using relaxation and imagery, only to be interrupted by the fact that my derriere, with its diminishing padding, was taking a pounding on the boat seat. The waves were getting very choppy. As the waves increased, the boat began pitching and yawing as we plowed along, slowing our progress even more. Soon I was willing to settle for just getting back without capsizing. But the truth was, none of this worked. I hate being late for appointments, and now I had two crazy patients waiting for me.

When we got to shore, luckily the Patrol car had waited for Tessier and me. The other doctors were tired and famished--not happy campers when I announced straight away I had to be dropped first at Notre Dame. But when I filled them in on the unfolding Red Cross drama they understood.

I led Tessier back through the hospital tents to our rendezvous with the Canadian Croix Rouge, catching Serge on the way out. "Yah, I know," he said. "We got a couple of hot ones for you. Brita's inside." She had chairs ready to go, and walked us over to this nice little bright white tent, just big enough for a desk, five chairs, with a little walk-around room, good for agitated patients and doctors with fanny fatigue. It was hot, but not by mobile clinic standards, so Tessier and I pulled out our water bottles, my bag of medication, my tush cush, pen and paper, and we began to wait.

Just a few minutes after four this attractive, clean cut, 20-year-old guy with glistening furtive eyes was led in by his sister, and a brother-in-law, both bright articulate people. He was jabbering away, his wild eyes dancing all over. When I tuned in, he was clearly looking at different things around the room; when he pointed upward, I asked him what he was seeing. He began telling me how Jesus was talking to him; the moon was really Jesus, and that the moonbeams shining down were Jesus speaking to him. Then his eyes turned glassy, he closed them repeatedly, scrunching them real tight, and shaking his head rapidly side to side, his lips and cheeks flapping out, spittle splashing like an excited Golden Retriever, as he sputtered, "Djab, Djab, the devils are everywhere, look at them staring in at me from the tent walls, oh, Jesus, Save me Lord." I thought we had our diagnosis, spoken loud and clear. He was clearly hallucinating and psychotic, with striking paranoid delusions. No pure delirium here. But what troubled me even so was the presence of any delirium. As we say in the profession, there was a smell of 'organicity' here, of something physical, something wrong with his brain, not just his mind. His press of speech and hyperactivity also spoke to a mood-driven manic-like state. But knowing I had to get a systematic history, I tried my Creole, and then my French which we all understood better. Tessier, bless his patient soul, bailed me out when he had to.

What we learned changed things dramatically. He had a history of a generalized seizure disorder (meaning total body convulsions), had run out of meds last week because of the quake, and had virtually gone in status epilepticus (almost constant seizures) the day and night before, having a string of nine seizures. This massive sustained electrical eruption of brain activity had driven him into a floridly manic psychotic state. At the same time, he had a post-ictal (post-seizure) acute brain syndrome (meaning new and reversible, causing the symptoms of delirium that concerned me). I couldn't do formal mental status testing because he was so disturbed, but on careful listening determined there was clearly delirium present, though dwarfed by his hallucinations, delusions, and manic state.

Okay, I said to myself, *the first order of business here is to quiet his underlying seizure disorder, but at the same time to cool down his psychotic mood disorder. Luckily, Carbamazepine does both. But he's so floridly psychotic he needs a nice 'wallop' of Haldol to help with his hallucinations and delusional thinking as well as the mania.* Pardon my doctor slang, but this was a tough situation, and this is the gallows counter-phobic way doctors think and talk. Wallop the mental enemy, win the battle. The earthquake had knocked this guy off meds and out of balance, and we had to do something pronto.

Normally a guy like this would be 'slammed' into an in-patient psyche ward, and 'jumped on' by staff. But we didn't have that luxury. Luckily, I had trained in a Harvard day hospital, and was used to sending these patients home if at all possible, provided they had a good family--which he had. So I upped the meds to an almost 'scary' level, though levels I was familiar with, intending to 'hit him hard' (albeit briefly) and then back off. I had a long serious talk with his family, explaining everything I was doing for him, and the risks of a dystonic reaction (neck spasm, for instance) drug side effect, then spoke directly to him about all this, trying to find that sliver of rational mind still present. Then I sent him home. But I gave them my mobile phone number and said I'd keep my phone on all night. Tom would love me if they called, but he would understand. At least I didn't go flying out of bed at the slightest earthquake. I also told them I would see them the very next day at 4 pm here at the hospital. This case required a very close follow up.

He had barely gotten outside the tent flap when my next patient, a 61-year-old woman, face wrinkled and eyes down cast, a blue and yellow bandana wound tightly around her head, was actually led in. Waddling because of a bad hip, all the while squirming and mumbling, she was pulled in by two of her grown children. She was so averse to seeing us she kept turning and pulling away, virtually coming in backwards. They had to push her down in the chair several times before she stayed put, sitting there sideways. Being Sigmund Freud, I realized she didn't want to see me, (not me personally since she didn't know me from Adam, but something I represented). Tessier had trouble deciphering her Creole mumblings. Her daughters explained her incessant mumbling was about how we were all going to die, how the earthquake was coming, how terrible it would be, how so many would be dead, about how terrifying it was.

Since the earthquake she had been in this state, not taking in any food or water unless forced, sleeping only briefly, waking up with fits and starts, averting her eyes, not wanting to look or talk, seemingly stuck, trapped in this state. I tried to talk to her, but she wouldn't look or listen, turning her back on me. Then she stood up, turned, and bolted out like a jackrabbit. Her daughter tackled her, but couldn't make her sit again. She complained of sharp supra-pubic pain, grimaced, pressing and rubbing her pubic area. Her agitation was the worst I had ever seen. But even in her constant motion, the particular way she squirmed, shifted her wait, and held her abdomen made me think of my daughter when she was little, couldn't speak yet, and had to pee. I asked her daughters, they walked her just outside, she squatted, and urinated for quite a while. With her scant water intake, I was impressed. She really needed to rid herself of something. Even so, this hardly calmed her intense restlessness.

Their house had collapsed, and there had been suspense around the survival of family members, but miraculously no one was actually hurt. Her husband was unharmed. We soon found out her mental eggshell had been cracked once before, and the experience of the earthquake had shattered her mind once again. Several years before, with the death of a sister, she had experienced a mental breakdown and exhibited the same behavior. What caught my ear was how frozen she was in time, seemingly stuck as if it were just before the quake, leaving her constantly predicting its horrendous impact for herself and everyone else unless we did something.

I felt I embodied the approach of the quake disaster. I told her, "I think you're trying to protect yourself, and everyone from the earthquake, by shutting everything out--not wanting to look, hear or think ahead." I also said, "I don't think you want to face me and hear any bad news, because you worry it will be about death of someone you love. By acting like it hasn't happened, you can warn and protect everyone." Her reaction made me feel like I had let a fart in a blizzard. My words were a seeming non-event. And yet I let her know I had taken in and acknowledged the blistering gale force of what she was facing. Having said my piece, I told her I didn't want to see her continue to suffer so, making it impossible to think and face reality, so I was going to give her some medication, first by shots to give quick relief, and then by pill. I had the helpful Red Cross Emergency Room doc draw up syringe fulls of Haldol and Diazepam, and while her daughters and I held her, she got two sizeable injections in her *derriere*. God knows what she thought we were doing back there, even though I whispered Creole clarifications in her ear the whole time. With Tessier's help, I went over possible Haldol side effects with her and her family, as well as the pill dosages for Haldol (for her stuck, reality-denying psychotic state), Diazepam (for her severe anxiety and agitated restlessness), and Carbamazepine (for her depressive incessant drivenness), along with careful instructions for management. The last things I gave were my cell phone number and an appointment the next day.

My shirt was wringing wet. I had forgotten about drinking my water, and my stomach was beginning to gnaw at me. Tessier was just shaking his head. He had never seen florid mental illness up close and personal, and these two cases were doozies. The Red Cross had found a couple of real winners for my initiation rite. He even went out and bought me a coke at the gate as we waited for the Patrol car, feeling I had earned it.

But the day was not over yet. We all arrived in time for dinner as the last rays of the sun filtered through the bars on the living room windows. Always a dramatic expectant moment, we lifted the fly protectors off, hoping to see a goat. But as usual we saw a chicken, and apparently a scrawny one at that. Tom said it was about enough for him, and that he would cook for us the next night. "I'll go out and get some real meat myself," he declared. We laughed and dug into our red rice and beans like good Haitian children--and shared medical war stories. My boat docs knew what I had been facing, and pumped me for details. My mine took the cake. They particularly liked my rendition of the young man shaking his face and lips as he was seeing devils on the tent walls. We do these things not to demean patients but to leaven the weight of what we otherwise carry privately inside, weighing down our souls. We need humor and laughter to rejuvenate ourselves.

Drs. Paul, Alice and Alisia donned their headlights, mixed their Clorox solution, and splashed into the dishes. They were letting me off to nurse my wounds. Instead, I began typing up my diary, feeling frustrated once again I had no way to send it to people because the Internet was down most of the time. Just then Stephanie came rolling in on a new bike, high as a kite. "Come look at my new toy!" It was a spectacular, shiny new, silver and chrome, well-sprung, dirt bike, 'Mongoose' boldly emblazoned on its cross member.

Stephanie with her new Mongoose bike outside the Petit Goave Residence

"Hey, guys, this baby can really take the worst potholes and bumps. Finally I'm going to have some fun and really get in shape!"

"When are you going to do that?" we said, almost in unison.

"Very early each day. I'll be leaving at 6 every morning. No one on the road. A good safe time to roll out of here for the Office. And my driver is ecstatic I'll be taking care of myself." We were impressed once again by our fearless leader.

Clearly in a good mood, she offered to chat with us. People started bringing up things, like the dim lighting, the cold water, the scant food, and finally the Internet, a pet peeve of mine. I don't know what got into me, maybe from my afternoon Red Cross workout, but I all of a sudden let loose on poor unsuspecting Stephanie once again, ranting about how I couldn't Skype my wife, send pictures or attachments, and half the time couldn't even read or send emails. As I warmed to my subject, I got sharper tongued.

Finally, Stephanie lost her patience, "Kent, I know the side of you coming out now, I've seen it before, as you know. But need I remind you how lucky we are to have any Internet at all any of the time. We are lucky to have half the stuff we have here, and none of it existed here just a few weeks ago. We're trying to house and feed, treat and cure these traumatized Haitians, protect you guys with costly security measures, and start up new clinics in Leogane and Jacmel, putting our precious scant resources where they're needed. Frankly, we are stretched thin both in support staff and money. And upgrading the Internet, at the moment, just isn't a top priority. We have starving mouths and sick patients to take care of, and you're complaining about hi-tech inconvenience in a virtual war zone."

I shut up fast, feeling greedy and guilty, and quite embarrassed. I like Stephanie and had shot my mouth off right after feeling so grateful to her. Though my issue was real and shared, I had put my foot in my mouth. As Patti has so often said to me, "It's not just what you say, Kent, it's how you say it. You get too hot and bothered." Only when I cooled off and removed my foot did I have a little talk with myself. It was then that I realized that she had used what I now call her 'Darfur Defense' on me, an all-purpose administrative counterattack covering a multitude of sins and oversights. Her Darfur Defense had certainly trumped my offensive moves, and that was okay, since it rang true enough to carry the day. But Stephanie did, even with all she was carrying and facing, have a tendency to put things off, to put out the newest brush fires (and forget about other doable things, things not on her priority list). She could be a little scattered, distractible, and impulsive. She also tended not to delegate much. Nevertheless, I had done enough administration, (and admittedly not in a crisis or war zone), to know how difficult and demanding it could be, so I had to hand it to her for her ability, perseverance, upbeat mood, and accomplishments. She rarely took time off, and almost never a vacation, though she talked about it. I even heard her talk about giving all this up and getting back to her life. But as soon as this came out of her mouth, the next thing she said was, "But if I did that I'd probably go sign up right away for an Arabic-speaking crisis zone. I love those countries." All told, though, Stephanie is a real winner--and I a bit of a hotheaded heel, at least today.

Thursday, March 25, Beatrice Clinic: Angel Trouble

Up bright and early, I was doing my exercises when Stephanie, biker shorts and all, came skipping down the Residence stairs. She grabbed her Mongoose and headed out the door. Feeling good and feisty, I decided to take matters in hand, cooking up 10 hard boiled eggs for everyone's lunch, and then whipping up scrumptious scrambled eggs for Tom and me (Jattu put her spoon in too), all ready and timed perfectly just as Crystal walked in late. My egg-shot across her bow began the latest skirmish in the daily Crystal wars. She let out a little sniffle and shuffled past me, a few bananas sticking out of her bag, and set about cleaning up last night's few remaining dishes. Two of us took a hardboiled egg, and the rest were apparently left untouched by the others. We had been asking for bananas and papayas for days. At least the bananas had arrived. We all made a verifiable banana sighting. But no one saw them again. The trail was getting warmer, only we didn't pay enough attention until comparing notes at dinner that evening.

Before my digression, we were on our way to Beatrice clinic, and soon we arrived. On the way I had a sad thought. This was my last visit to Beatrice. Next week this time I would be leaving for Port-au-Prince, the day before my departure home. Time was flying by. It was then I realized we had a problem. How would all my patients on psychotropic medications be getting theirs? I had all the meds in the red backpack I toted around. We kept most of the psych meds separate from the other medicine handled by the pharmacy. I had to devise a plan for continuity of care. More worrisome, who would be taking my place to cover the teaching and clinical care? I had to get my head together and make some calls fast. But just in case, I hit on the idea of leaving enough medication with each Clinic staff to cover a couple of weeks, in case there was a gap.

I couldn't believe this hadn't crossed my mind before. Although in all fairness, I was so consumed by starting up the mental health team and preparing lectures that finishing up was the last thing on my mind. And once I got rolling, the cases and teaching were so demanding I was hard pressed to look beyond the next day. At least Lynne had let me off the hook around coming back in to Port-au-Prince to do the Mars and Kline lectures.

And what about IMC? As the parent organization, where were they on all of this? One thing I did know was that, even as they were creating new mobile medical clinics, for instance in Jeremie and Leogane, they were also in transition from the acute care phase of their Haiti effort, staffed in large part by volunteers coming for just two weeks, to a more sustainable, longer term effort, staffed by more permanent paid people and long-term volunteers. But what would be happening here in Petit Goave? I was a volunteer, but maybe I would be replaced by a paid staff member who would stay longer. I would have to ask right away.

At the Clinic that morning the first two patients had had their symptoms long before the earthquake, though the quake and 'repliques' didn't help any. One had dizziness and headaches and the other trembling of his hands. The 54-year-old woman had been to the hospital OPD, and got some medicine relief, and wanted a refill. We gave her what she needed, and in addition, some stress intervention techniques, after going over her psychophysiology with her. The 61-year-old man wasn't the only member of his family with trembling. I leaned over and whispered to Dr. Polo, "Have you noticed my hands? They tremble too." "

"Yes, I didn't want to mention it, thought maybe you were just excited." He was being kind.

"He has what I have, only worse." He and I discussed the condition and went on to explain hereditary familial tremors, acknowledging that his excessive alcohol use improved the tremors in the short run, but his continued heavy drinking would make them worse in the long run. Even so, earthquake stress can make such tremors worse. Because he also heard voices and had significant sleep difficulties, and was somewhat loose in his associations, Dr. Polo and I decided to 'tighten up' his thinking and provide some sleep relief with a small amount of Chlorpromazine. When we asked if he had seen a native healer, he surprised us, revealing that he was a Voodoo Houngan himself, but was retired. He went on to tell us he had always had faith in modern medicine, and was good at referring when indicated.

I stepped outside to stretch, but really to enjoy the breath-taking view again of the Petit Goave Harbor below, just beyond the valley and outskirts of town. It was high noon. The two Tankers, rusty red and dirty white, were still moored to the far left. *Shipping must be at a stand still,* I thought. I heard a commotion down the street. A man in a straw hat was shaking his fist at a woman under a green umbrella selling papayas, pointing at another papaya stand across the street. The woman held her ground, her voice sounding angry. The discord bothered me as they exchanged heated conversation, but I waited. Finally she put one hand up, raised a few fingers, and they had a deal. Both began laughing. If you didn't bargain energetically in Haiti, you weren't any good at business, and you weren't having fun.

Walking past them was a parade of stately women, balancing their wares on their heads, coming back up from Petit Goave market. The younger girls, many beautiful and statuesque, glided past me, their breasts jouncing, their hips swaying beguilingly. I was noticing something new. I had been so stressed and anxious during the first weeks in Haiti that women were the last thing on my mind. But now the sap was rising again as I calmed down and hit my stride, and it was nice to feel more alive again.

As I watched I remembered why I had found Haiti so tantalizing and seductive when I was 21. Back then it didn't take much to make young sap rise. Many moons had passed, but I smiled remembering a certain Haitian girl, the first girl I ever had, a beauty I met on a bus trip to Les Cayes, far out on the southern peninsula. It had been a long bouncy fascinating ride. We happened to be sitting next to each other, she going home to see family. We kept jouncing into each other, soon laughing and enjoying the physical jostling. By Les Cayes, we had gotten to know each other quite well. We didn't want it to end, and so we had a cup of coffee--an innocent beginning to a first tango in Haiti. Ah, sweet bird of youth! As I reminisced, a particularly beautiful girl walked by. "Bonjour," I said. She smiled but looked at me quizzically. I thought *it must be because I'm a Blanc, or maybe an old guy.*

But Tessier, standing behind me by now, laughed. "It's after 12 now. You're supposed to say 'Bonsoir', not "bonjour! Remember?" In Haiti, for some reason, the evening begins just after noon, maybe because they get up so early. Haitians are pulled into the rhythm of the sun and moon. The mornings are cool, the evenings dark and often lightless. But Tessier knew something else, that we guys shared the enjoyment of the way she walked. It reminded my of that old Livingston Taylor song, "I love the way she walks." I had worked on the psyche ward he had been on briefly, and got to know the beautiful head nurse he wrote the song about. She was strikingly beautiful, but that wasn't the story. It was the incredible way she moved. Tessier and I were sharing a Livingston Taylor moment together--very special. But we had patients to see.

We walked back to our Clinic, the large tent city sprawling out behind our olive tent, climbing up the mountainside. *Many patients come from those tents,* I thought, *but many more walk miles down out of the mountains to see us. They deserve our attention.*

Our last three patients all had seizure disorders. The run on seizures was continuing. As I thought about this run of seizures, and the decrease in acute cases I was seeing, I realized I was experiencing tangible clinical proof of why IMC was shifting gears into sub-acute phase of longer term staff planning and clinical care. They knew what they were doing. They just hadn't told me about it and their transition plans for Petit Goave yet.

We saw a 33-year-old man, looking very sad, who had one or two seizures a month. He knew when they were coming because of his telltale aura. But they happened so infrequently he hadn't gotten treatment. He finally admitted he felt too embarrassed by them to tell a doctor. We began him on Carbamazepine, discussing the demoralizing effect of constant worry about their unpredictability and the insidious humiliation, emphasizing regular medicine compliance would be crucial to his success.

A 16-year-old girl walked in with frequent seizures, each heralded by a sudden turning of her head and eyes. She told us they were accompanied by a pervading sense she was seeing the devil getting closer. There were hints of a thought disorder and mood fluctuation, but she had just seized that morning. Though we were tempted to think of incipient psychosis, Dr. Polo and I elected to give her Carbamazepine and some initial Diazapam, realizing post-seizure thinking was often a little loose because of residual brain irritability. We wanted to be careful, using follow-up to track her closely to make sure her post-ictal acute brain syndrome abated fully. We always worried about missing some other process going on.

Our last patient, a 20-year-old, had had seizures about once a month since age three, but because others in her family had them, she was advised they were hereditary, or maybe an ancestral curse, and so her family felt she shouldn't be treated. Who gave this advice? A family *Mambo*, or female Voodoo priest. The priest was less medically astute than our previous *Houngan* patient. When we said we would be happy to go ahead and treat her seizures, she said, "Thank god. I've been down on myself all my life, sometimes thinking of killing myself. What a relief." We were pleased to begin her on Carbamazapine, but would need to have follow-up visits. I guess the word was out we could help seizures, given the run this morning. Then I added, "We could consult and update your family Mambo so she won't be upset, and you could tell the other family members with seizures to come in and get help.' She looked at her feet, and said, "She died in the earthquake. That's why I came in. They'll come in now, too."

As we finished our last patient, I realized our facial burn patient hadn't come back for her seizure follow-up visit. I mentioned this to the nurses and they said they would put the word out for me. I made a mental note to email Fred Stoddard again at Boston Shriner's.

During Beatrice Clinic, Brita had called saying she had a meeting but would try to get there before our Red Cross patients. Then she said, "Uh, Kent, there may be another one from the camps, pretty depressed. We're worried about suicide."

"Okay, put her last, about 5:30, since we'll move faster with the first two because they're follow-ups."

It was pushing three by the time we finished. Tessier already knew where we were going—to our new Notre Dame Clinic. When we arrived, Brita was nowhere to be seen.

We went into the Red Cross office to scrounge up some chairs. Eyebrows raised as we walked in, then someone said, "Oh, wait a minute. Brita mentioned you were coming, her meeting is running over, and the other patient is doing better and refused to come in." My feelings weren't hurt at all. And then, to my surprise, he got up and gave me his very own chair, plus two others. We were getting a real welcome.

Our 20-year-old came in following his brother-in-law, still looking at the tent walls suspiciously. He was moving a little stiffly.

"Uh, oh," I said. Tessier looked at me quizzically. "See the mechanical way he's moving? That's called Parkinsonism, a side effect of the Haldol." But he was sleeping better and had started to eat.

"Jesus is still talking to me, but he's helping me beat the devils. Look, there aren't so many on your tent walls." Then I saw him glancing out the tent flap, raised for ventilation.

"What's out there?" I asked.

"That's where they are. See them staring in? But they're scared of us."

"Oh, well, they better stand back farther," I said. "We're going to help even more. I'm going to keep your pills the same, but add one to help with your stiffness and walking." We added Kemadrin. In general, his brother-in-law was encouraged. Things were heading in the right direction. "We'll see you tomorrow." We had decided not to reduce his meds yet, even with the side effect, not until the devils decided to hightail it.

The family with the agitated woman was nowhere to be seen. It got later and later. I made a call and they said they were at the Notre Dame gate. I was relieved. Stepping out, I saw them walking along. They were moving slowly, but she was walking under her own power, though still squirming some and favoring her leg. As she caught sight of me, she turned her head away a little, averting her eyes, and slowed her pace, but moved forward still. Only at the tent flap did they have to grab her hand to get her in. She preferred to stand. "Let her, if she wants to, but we'll talk a while, and that chair would be more comfortable. She sat down reluctantly. Her agitation was diminished, her tummy didn't hurt as much, and her muscle aches had gone away. She was taking in food and water without much coaxing.

"You must have really given her some medicine. She's been sleeping a good part of the time since you saw her, and we've been able to also." Though her meds were high, we decided to keep them the same for at least another day, and were pleased she showed no signs of stiffness. They were happy to come in again right away. All in all, our Notre Dame Clinic was a success—so far. We picked up the chairs, and stowed them safely inside the ER door. I was ready for my coke at the gate, and bought Tessier one.

When I got to the Residence, there was a surprise waiting for me. And she looked like a Vogue model. She was 24, petite and demure. And I already knew her—the Haitian nurse Peter and I had interviewed for the psychosocial position on my team. I thought he had kept her for Port-au-Prince.

"Hi, Dr. Kent."

"Hi, Nathalie, great to see you." I was more than surprised. I was a bit anxious and miffed. *Thanks for the great advanced warning, guys. Now what am I supposed to do with her?* But I knew she was good from that initial interview and I was certainly 'more settled in'. The other message waiting for me was that Dr. Nick would be coming tomorrow night for 2 days. Now I had three reasons to call Nick: what was her nursing role, what were his plans for my team's transition, and how would we handle patient medication management? With all this brewing, I was glad I had already finished preparations for my Saturday Seminar.

But the first order of business was dinner. We didn't even have to look. Measly little bits of chicken floating in red palm sauce, red beans and rice. There was one surprise. Crystal had finally filled my request for fried plantain, but only 7 pieces. Now we had eight people at dinner. I heard Tom rummaging around in the kitchen, and something sizzling. It smelled good. Turns out he had taken an early peek and gone out to buy more chicken and other goodies. He was busy at the stove. We ate like kings.

"Hey, guys," I said. "How'd you like those hard boiled eggs I made for you this morning?"

"What hard boiled eggs? We didn't get any." Only 2 of the 10 eggs could be accounted for.

"How about the bananas I saw Crystal carrying in? I sure didn't get any."

"Neither did we." Jattu added, "I stopped by briefly around noon. Crystal was feeding chicken to one of the drivers, and gave him some bananas." The plot was thickening. Nathalie sat there quietly eating petite portions using perfect European table manners.

Then the Mongoose appeared, pushed in by a panting Stephanie. "Hi, Guys. Oh, and welcome, Nathalie. Enjoying your dinner?" Nathalie nodded her head. Some of us were about to speak up (not me), when Stephanie, trotting up the stairs, saying, "I have to shower. I'm late for a meeting."

I spent the rest of the evening filling Nathalie in on the clinics and how she could help. As a nurse she could handle and dispense medicine. I even envisioned her serving as the medication bridge between me and whoever was coming next. I was beginning to feel an angel had appeared. As the evening wore on, my anxiety melted. It was good to have someone to plan with, even if she was the same age as my daughter.

I finally reached Nick by phone. He said he'd be arriving the next night, stay over for my seminar and meet all the doctors and nurses. He was especially interested in getting briefed on all the clinics. "But Nick," I said, "You haven't answered the big question. Who's covering when I leave?

"I will be, but only on a half-time basis."

"How are you going to do all this half-time?"

"That's what we need to figure out."

I realized we had our work cut out for us. "What about Nathalie? She just arrived."

"Just take her around with you. Orient her. She can help with medication, maybe do some of the follow-up and run some groups. We hope to have her find some community volunteers to help with patients and outreach. Oh, and remind Stephanie to find me a place to sleep."

Friday, March 26, Chez Les Soeur: Musical Beds

When we appeared at the Office the next morning, the drivers, to a man, took a look at Nathalie and smiled. She was again dressed immaculately, with a stylish, somewhat low-cut dress, and smart new shoes--definitely a cut above what everyone else was wearing for disaster clinic work. I introduced her to Tessier, who responded warmly, glad to have a psychosocial nurse joining us. Joanne was my next introduction. Jattu had already had the pleasure the night before. We hopped into the patrol car bound for Chez Les Soeur, an extra chair tucked under my arm.

Working with Dr. Guirlande this morning, we started with a 7-year-old boy upsetting his mother by beating his head against a bench every day to the point that he now had frontal bony bossing (a prominent bony bump) on his forehead. He had been retarded since hospitalized as a toddler for typhoid and malaria. More recently he developed spells, complaining of a smell, hearing a voice, and running, running, running—but no convulsions, just several hours of memory loss and sleepiness afterwards

"Sounds like psychomotor seizures. What do you think?" I said.

"But the voice and smell, aren't those hallucinations?" asked Dr. Guirlande.

"Right, but in a context, in a sequence, the 'psycho' part of psychomotor seizures."

"Oh, yeh. That makes sense."

"We can't do much about the mental retardation, except hope some of it is from ongoing subclinical seizure activity. We can bring that under much better control. Then we can see about the retardation. We can always hope for the best. No promises, but hope." It was nice to turn to Nathalie at that point, handing her the Carbamazepine, so she could count and dole out the pills, nicely placed in a marked plastic sac. She was quick, exact, and neat, handing it to mother with clear instructions spoken in Creole. Tessier and I were off the medication hook.

Our next patient came in with her aunt. She was a 22-year-old girl holding her baby who had trouble speaking because of a tight, tremulous throat that ached. "Feels like a knot in there," she said. She also had dizziness, hyperventilation and trouble keeping her eyes open when her throat was aching. This had happened twice before—once after failing premed exams, and again after failing social work exams. After the earthquake, when her favorite brother was crushed by their collapsing house, spells became much worse, now coming in waves. The moment she realized he was buried alive, she couldn't open her eyes for hours. Her other brother in Cap Haitien kept calling and crying on the phone, but she couldn't shed a tear. Tessier tapped me on the shoulder, "Dr. Kent, I knew her brother, a student and a friend of mine. He was a great guy. It makes me all sad inside hearing he died." I noticed a rim of tears in his eyes.

"Tessier," I said. "Tell her what you just told me."

When he did, she initially began to cry, then her eyes scrunched shut and her voice tightened, sobs catching in her throat. "That aching lump has come back," she said.

"Dr. Guirlande," I whispered, "she just developed her throat symptom, called a *globus hystericus,* right in front of your eyes, an anxiety-based laryngo(throat)spasm. Some people call it a 'stifled cry'. These symptoms are from her intolerable emotional loss, blocked by her mind, forced to spill over into her body. She wants to close her eyes to the painful reality. When her normal waves of sadness hit her, her defenses block them through what are called conversion reactions resulting in her symptoms."

"But I have a hunch we're still missing something. Ask her what was going on with her and her brother. Something is making it hard for her to mourn. One of my best supervisors, Hyatt Williams, always reminded me, *Kent, the devil is in the details.* "If we can loosen up her arrested grief reaction we can put her on the road to recovery."

When Dr. Guirlande asked, she found out that the patient and her brother had just had a big fight, and she was really mad at him.

"Mad enough to wish something terrible would happen to him?" I said.

"How did you know? That's why I ran out of the house with my baby, leaving him alone in there just before the earthquake struck. Why him and not me?"

"You feel horribly guilty, don't you, every time you start to cry, and you can't bear it."

"It's awful."

"Well, the earthquake wasn't your fault," I said, "nor your survival, and now that you've finally told someone I think you'll be able to cry for him when you need to."

I then said to Dr. Guirlande, "We need to get her family to help her face her brother's loss and tolerate her mourning."

"Should I just say this to her?" said Dr. Guirlande.

"By all means, but also include her mother standing there." We soon discovered she was actually our next patient. She had headaches, stomachaches, sleep difficulties, and feet that felt heavy. She had started to cry as she listened to us with her daughter. After hearing her out, we prescribed conjoint mourning, asking the daughter to teach her mother some of the sac, relaxation and imagery techniques. We prescribed joint calls to her brother, asking the daughter to take the initiative, organizing and dialing the calls and bringing up his death. We wanted to turn her passive stance into active participation. "Have her come back to see you briefly for several follow-ups. You can really help her a lot with very little."

Nathalie and I then made provisions for the Chez Les Soeurs doctors to have a two-week supply of meds for our shared patient follow-ups to tide them over the transition between IMC psychiatrists. I was well aware that doing this would make staff aware of my leaving, and the approaching transition, giving us a chance to discuss it.

Because Nathalie spoke French and Creole, and pretty good English, I asked her and Tessier if they thought she could handle Notre Dame Clinic. If so, and he didn't mind, he didn't need to come. He thought it was okay, and had something he wanted to do anyway. On the drive over, I briefed Nathalie on our two cases. She had worked on a psych ward for a month as a nursing student. She spoke highly of Mars and Kline again. I told her that after introducing her, I would let her talk with the first young man, to give her practice, let her catch up with his history, and see how he reacted to a woman.

When we got back to the Red Cross area I was shocked. Our nice white tent was gone. Trotting over to the Red Cross office, we met Brita coming out, who pointed to her right. There it was. "Come and see," she said. "I have a surprise for you." There in the tent were the desk, all the chairs, AND a fan, purring away sending a nice breeze our way. We felt our status had been nicely upgraded again. I didn't want our patient to be as shocked as I was, so I periodically cruised over to see if they were coming, catching them just as they were noticing the disappearing tent. Our patient was all smiles, and took to Nathalie right away. After her interview, his brother-in-law said, "He really likes Nathalie. He has a different gleam in his eye." Later, Nathalie told me she had problems with a couple of patients like this, one actually stalking her. She found it uncomfortable.

From her interview report, it became clear he was much improved, the hallucinations virtually gone, no seizures, and much more coherent speech. He still believed Jesus was talking to him through the moon, but didn't hear the voices or see the devils. His rigidity wasn't worse. So we cut his Haldol back considerably, and kept his Carbamazepine and Kemadrin steady. I felt he was doing well enough he could go over the weekend to Monday, after Petit Guinee Clinic. I let them know that the following Wednesday would be my last time with them, preparing them for a transition to Dr. Nick. Nathalie organized and gave the medicines in separate sacs, explaining each carefully, and reiterating that they could call me if anything came up over this lengthier break. Since they came a long way, this was welcome relief for them. Our young patient said a particularly warm goodbye to Nathalie, who shook his hand firmly, and tried to pull it away quickly. He held on a bit long. Afterward, she told me it really bothered her, and admitted this kind of thing happened to her too often with younger men. She said one even stalked her for a while.

I noticed out of the corner of my eye that our next patient, the 61-year-old agitated woman, was standing near the old tent site, looking around. I went over, said hello, and she actually looked me in the eye for the first time. I brought them over, introduced Nathalie, and left her to talk with them while I went over to find the ER chart. She found out that the patient was sleeping and eating much better, was muttering less, and beginning to converse more normally. But she was still somewhat restless and squirmy. Her agitated psychotic depression seemed to be getting better slowly. We reduced her Haldol to a much lower dose, keeping her Carbamazepine at the same level. We both felt comfortable having her come back the following Monday afternoon for my second to last clinic, explaining that Dr. Nick would be taking over after that, but that Nathalie would be there still. The family felt this was fine.

When the driver delivered us back to the Office, Dr. Nick was sitting there waiting. I introduced Nathalie, and suggested she fill Nick in on our two Notre Dame patients, since he would be picking them up for direct treatment when he came back. I went off to do some photocopying for my lectures the next day.

When we arrived at the Residence, Nick, a bit bushed, reclined on Dr. Paul's famous 'spider' couch. We regaled Nick with Residence stories, until we mentioned the descending spider. He sat up rather suddenly, looking around, then bedded down again comfortably. Nick is remarkably relaxed under most all circumstances. He seemed to doze off just before Stephanie came in.

The whole time we were talking I was wondering where Nick would actually sleep. Because Nathalie arrived, Dr. Paul, our chivalrous doctor, offered to vacate his upstairs bed for her, pitching his pup tent as I mentioned earlier, right in the entranceway. But now we had an additional problem with Nick. Since Stephanie had recently lost a roommate, who had slept on a mattress in her room, we all thought that would be a natural solution for Nathalie to go there.

However, when Stephanie arrived, Mongoose in tow, she took surprising exception to this plan. "Slow up, guys, not so fast with my privacy. I haven't had much companionship lately, but something is happening, and I am going to need my space. Working the way I do, there isn't much time or chance for romance, so I have to say no for now. Nick, we'll figure out something I'm sure." A few people had quizzical looks on their faces as she rushed up the stairs, but I knew she was talking about her Oxfam boyfriend. She was hoping for an intimate moment, finally caring for herself a little. I was pleased.

"Oh, don't worry about me," said Nick. "I can sleep right here on this couch. Feels pretty comfortable, and the spider will keep the mosquitoes away."

We all laughed, and knew Nick could handle it. Stephanie went upstairs for only a moment before she called Nathalie up, telling her she could sleep in her room, since Nick was only there one night. Nathalie, bless her soul, declined, saying she would rather sleep out front in a pup tent. So the problem was solved. I figured when Nick came back, he could have my spot. I would warn him about the dangers of rolling toward the center of the bed.

Because it was Nick's first and only night, everyone wanted to head to the Royal for some lobsters. Nathalie didn't have the money, so Nick and I sprang for her supper. We couldn't leave her home alone on one of her first nights. She had lambi, while Nick and I did the lobster splurge. As a result my coffers dwindled to zilch, mainly because of my big contribution to Pierre d'Haiti (and previous lobsters). So I would have to skimp from now on. I knew I'd get the second half of my per diem back in Port-au-Prince, but how would I finance my final Barbancourt and lobster next Wednesday? Maybe Stephanie would give me an advance? I had to keep my mouth shut and be on good behavior.

I turned in early, wanting to get my beauty rest for my seminar the next day at the Royal. Luckily there were no repliques. Tom and I slept like babies.

Saturday-Sunday, March 27-28: Seminar and Salvation Army

The previous Saturday seminar the doctors and nurses asked for handouts, in French and Creole, and I had them ready. They also wanted more teaching activities, and I was loaded for bear. They wouldn't have much fanny fatigue. I also had caught on to how Haitians manage seminar time schedules. Starting at 10 meant 10:20, and breaks were run on 'Haitian time', not American time. So I trimmed everything down to these realities. Drs. Alice and Judson had a little presentation on vaginal infection and diagnosing child sexual abuse, and asked if they could have 20 minutes. I said, "Sure, take from 10-10:20." Wasn't I generous? I figured I'd let these two eager beavers with their hot topic draw people into the room for us. It went to 10:30, but I was expecting that. Alice and Dr. Judson were a great act together.

After introducing Dr. Nick and Natalie, I launched into my topic with more comfort and volume than last time, knowing that they needed techniques and juicy vignettes around doing the Psychiatric and Mental Status Exam. Tessier upped his volume some, though he still needed to be louder. He was used to children, acute of hearing, not deaf distracted doctors. I hit them hard with some material, and then had them pair up and do the mental status exam on each other, switching examiner-examinee roles. I then asked for volunteers to describe both sides of their experience, calling on people directly if no one responded. They were in for trouble now because I knew their names.

I was also well aware why we always had such a good turnout for these seminars. The Royal threw a great buffet. Because it was Dr. Alice's last full day, she was going around to all the tables, saying goodbye to the Docs and nurses she had worked with, and trading email addresses. As I watched I felt a particular sadness. I really liked Alice a lot. We allotted plenty of time for the luncheon. Getting them back in was like herding turtles. But I had announced we had a surprise in store, so they came along.

Dr. Nick was on stage with a very interesting activity allowing everyone to experience auditory hallucinations. Using me as the guinea pig patient, he had another doctor try to interview me while he was speaking paranoid thoughts into my ear. I was distracted and confused, actually becoming suspicious of the doctor. It jammed my thinking and was quite crazy making. Then he had everyone do it. The exercise was a great teaching success. After a long Haitian break, I lectured on the cardinal symptoms of the major psychiatric conditions, sometimes talking in English, sometimes in French, and even in Creole. It confused Tessier, who was patient, and amused my audience. It was my swan song so I hammed it up a little, keeping their attention. Finally, I had lined up several of the doctors to give case presentations from our clinics to illustrate these conditions. They did a great job, including Dr. Affricot acting like he was President Preval of Haiti at one point, carrying on as if he were talking with Sarkozy. The evaluations at the end gave me better marks, and I was pleased—and relieved.

I was looking forward to Crystal's Saturday night dinner, since she had promised goat meat for the first time. But first, I headed over to the bar for my reward, a nice double shot of Barbancourt, and a tall coke. *Cuba Libre*, if I had a lemon. As I sat there celebrating alone, Paul and Alisia showed up, sliding in next to me for a Heinekens. Tom was rubbing off on them. They had just had an emergency call from Stephanie and were celebrating helping her out around another little girl. A neighbor's child had fallen and broken her arm. Stephanie called this ready doctor team and they popped over to the rescue. My kind of Docs!

As I listened, they mentioned they had noticed Stephanie's face had a big ugly scrape mark. When they asked her, she launched into Mongoose stories about potholes and her chain derailing. She was stuck on the road with her dead Mongoose and this guy stopped to help. He turned out to be the very businessman she had just had a big fight with. To her surprise, he was gracious and glad to help. She was impressed. "Chivalry rides again," said Dr. Paul. But Alisia wouldn't be thrown off the trail. Stephanie hadn't really explained her facial abrasion. So she pushed Stephanie.

"Okay, Okay," Stephanie confessed, "I got it from brushing up against some coral. Let's leave it at that."

Both Paul and Alisia became concerned because coral wounds can get quite infected. But I was imagining Stephanie out skinny-dipping by moonlight with her OXFAM boyfriend, the one stirring her wish for privacy. Though I was curious, I also wanted a loan, so I held off teasing Stephanie. She would have to remain a woman of mystery.

I couldn't wait to get back to the house and have some of that goat. Dr. Paul was up for it, too. On his calls home to his wife, she kept asking about goat, but he was beginning to despair. As he put it, Crystal's finally getting *my* goat. Well, tonight was the night, last chance. Forks on the ready, we whisked away the fly protectors, and, low and behold, there indeed was some sort of mystery meat, definitely not chicken. But there was precious little of it. "All for you, Paul," I said. "Oh, no, I wouldn't know goat if I tasted it. Somebody else has to make the call." It was cooked to grey kernels, might well have been goat, I couldn't be sure. What I did know was Stephanie had given her plenty of dough to get lots of goat. Fifty years ago Joselia, Ternvil's wife, could have bought me an entire goat for that money, enough for her whole family and me, taking inflation into account. Something was off here. 'One thing is clear," I said. "We're not getting what Stephanie pays for."

Jattu cleared her throat, "I've been around when she leaves a couple of times this week. She always leaves carrying big bundles, more than I saw her come in with when we were exercising, Dr. Kent." Tom was already in the kitchen cooking spaghetti.

We all looked at each other, and light bulbs started going off. "She's stealing, and starving us!" we said in unison, the dawn finally breaking over our numbskulls.

"She must be feeding an army at our expense." "And courting the drivers on the side. Stephanie will be furious!" I had a private thought later on, *Actually, Crystal might in fact be feeding a lot of family members, people who would otherwise go hungry. Nothing is ever simple in Haiti, is it? But IMC has its rules, we have generous defined ways of giving, have a right to be fed, and this needs to get sorted out.* But who would be the whistler blower? I was a Raven, not a stool pigeon. And I needed Stephanie on my side.

My last Sunday was approaching, but it would be no beach day, no picnic. I was in the home stretch and had all my medical chart work and statistics to do, trying to pull things together so I could pass them on to Nick. I also knew they represented my contribution to IMC research projects. Could I at least sleep in a little? Well, the good lord had another idea for me. I began to go to church in my dreams, at least at first, then I actually woke up with a start. My watch told me it was 8 am. The Pentecostal church service was in full swing next door—and I mean swing. The music was rocking. And what sermons! Though the language was different, I knew the cadence. I had been awakened by some fabulous singing followed by a pretty bombastic hell-fire sermon. I knew all this by heart and soul. I had attended Pentecostal services my junior year in New Haven, nearly undergoing total immersion baptism at Beulah Heights First Pentecostal Church. It had started out as my Culture and Behavior field project, but even though I switched to Haiti I kept going, deeply moved by the people, the music, and the message—a very different experience from my Episcopal upbringing.

I sat up straight, feeling guilty about how bad I had been about my record keeping, and what I needed to accomplish for redemption. We cooked our own eggs and headed over to the Royal, where the ambiance was better and they had big tables where we could spread out our work.

We chose a spot that happened to be right outside the room of a newly arrived camp nurse from the Salvation Army. She was buxom, eager and almost attractive. When I heard her describe the camp clinic she was facing, I asked Dr. Paul to orient her. She found this very helpful, relieving pressing questions and her start-up anxiety. So she was exceedingly grateful. Paul, who had been practicing his card tricks all morning, asked if he could use her toilet. He didn't come back for the longest time. She finally went in and found him dead asleep on her bed. She let sleeping chivalry lie and came back out. My Mac ran out of juice, and she let me plug into her room socket, careful not to awaken Paul. Though I had my tush cush, I eventually got fanny fatigue--and finally a back spasm. Walking it off was the only solution. She asked why I was parading around rubbing my back. When I said 'back spasm', she said, "Oooh! I'm a good nurse. Can I rub it for you, I'm great with backrubs. Ooops. Maybe I shouldn't have said that." I considered my options and decided there was no fire under my snowy roof at the moment. I felt certain my salvation lay in doing my statistics-- not other figures. I said, "Why don't you go in and wake up Dr. Paul? He might have a few kinks to work out." She seemed intrigued, since he had already been so helpful. But I knew full well Paul would be the last one to get kinky.

After while, I needed to stretch my legs again, so I walked off around the grounds, ending up down near the little entrance bridge. The moment is burnt into my aging brain. It was near some half buried colonial canons and tents. Lost in thought, I heard my name called out by a familiar voice. There was Crystal Wells again. Someone was following her—Sienna Miller, IMC's Global Ambassador. This time I didn't bury my head like a dodo. Perking up properly I said, "Hi, Crystal. Hi, Sienna. Nice to see you two again."

Sienna walked right up to me and pulled something out of her pocket—a little hand-written note—and placed it in my palm, as if she had been waiting to see me. "I saw this depressed woman in Platon who really needs your help. Here's her name. By the way, they like what you're doing out there."

Slipping it in my wallet, I said, "Thanks. I won't forget it." And with that they walked on by, leaving me standing on the bridge—an old Billy Goat Gruff.

Monday, March 29. Petit Guinee: SNAFU

Monday rolled around all too soon. I was sad about doing my last clinics, not even a full cycle because of leaving Wednesday. My clinic that day would be Petit Guinee. And my statistics weren't quite done. It was down to the wire. Back at the Office, Nathalie, Tessier and I grabbed our chairs and a folding table to head for Petit Guinee, our last teaching session there. When we arrived we were greeted with the first of several surprises. There was a second huge white tent set up next to the first, completely empty. "Finally we have our own tent," I said, laughing. We went in and set up our office in the middle. After a while, a spunky new nurse came trotting in, telling us the tent was for the Nutrition and Nursing Mother Program. While we were talking, several mothers came in, made themselves at home in our chairs, and began nursing contentedly. We had been displaced, and felt too guilty to take our chairs back at the moment. When we went next door to the main clinic tent, I was surprised to find only one Haitian doctor, and not the one I was expecting to work with.

"What's up?" I said.

"Oh, didn't Joanne tell you, the other doctor quit and we had to move things around. Sorry."

"So my last teaching clinic is SNAFU," I groaned. "I'm really disappointed, especially after my teaching seminar."

"What's a SNAFU?" said the Haitian nurse.

In English, I said, "Situation Normal All Fucked Up." She shook her head, Nathalie looked quizzical, and Tessier laughed.

"Let's call Stephanie," I said, "and get some new marching orders. She's our trouble shooter."

She thought maybe Chez les Soeur, where they had two doctors, but it turned out to be the first day for our physician-pharmacist freshly minted and pressed into front line service there. Not a good time to teach him, or the other doctor, leaving a new doctor holding down the fort. So we pulled up stakes for a second time, and headed back to the Residence. Tessier went on into town. I was feeling at a loss, but then I remembered something I knew was coming later in the day. I reminded Nathalie that we had a special Cluster meeting to attend along with Stephanie that afternoon at 2 pm—the very regional NGO Cluster Meeting that I had thought so boring my first week there. Now it was the most interesting thing on the horizon.

When we got to the Residence, we had some time to kill, and I could do my statistics, but Nathalie asked if we could talk first. I said sure. We sat down, and out came all the conflicts and heartaches she was having with her orthopedic resident boyfriend back in Port-au-Prince. She surprised me, getting into quite a lot of intimate detail.

"I'm very flattered you'd share all this with me, and I have a few thoughts. But why now?"

"Dr. Kent, I really like the way you've been talking with patients, and helping them. And I haven't had a father since I was 10."

"Since 10?"

"Yes, that's when he died."

"Died so young, what happened?"

"We don't know for sure, but he went down to the beach one morning to swim. He was a champion swimmer. We were all quite comfortable in the water, and I loved to go swimming with him. The next thing we knew, they came screaming up to the house saying he had drowned. When we got there, he was in 3 inches of water, barely breathing. He died on the way to the hospital."

"I'm so sorry for you. And you've had to spend so much of your life without him. Has your mother remarried?"

"No. She's had some serious relationships, but no. She's a great mother, but it's not the same. I really miss him."

"I would, too. It's very sad, but more, missing out on having him for so many things during your life growing up."

"And the way you've been talking to people about the loved one's they've lost, that reminded me of him."

From what she said, I had a few ideas, and felt she really needed a sympathetic ear. Once again I felt like I was with my daughter. As a psychiatrist I had learned to empathize not just with patients, but with everyone a patient talked about, putting myself in everyone's shoes, not just the patient's. With Nathalie, that included her father, but also her boyfriend. For an aging gentleman dealing with waning sexuality, yet still tantalized by an active fantasy life, thinking about being her boyfriend was hard on me, less so being her father. The back of a man's mind doesn't know the word 'no', if one is honest, even as the head and the heart say 'no'. These are the kind of thoughts one doesn't often share, not even with oneself, unless one is a shrink. One savors and guiltily dismisses them. But I tried to keep them tucked away in mind. Something about Nathalie was percolating inside me, something I hadn't formulated yet, and I needed all my memory marbles available to figure it out.

Just as the 2 o'clock Cluster meeting approached, I got a frantic call from Stephanie wanting to make sure I was going. It turned out she couldn't make it, something about a baby being dumped in her lap and needing to deal with it and some urgent legal matters.

"I need you to represent IMC at the meeting."

"Sure. No problem."

But when I got there with Nathalie I soon realized there was a big problem. The meeting was entirely in French, and the Royal terrace was filled to overflowing with about 30 NGOs from the Petit Goave region, all dealing with psychosocial issues. The meeting was run by the Canadian Red Cross, luckily backed up by Serge from Notre Dame—a familiar face. On the way in, he asked if I wanted a ride back to the Hospital after the meeting. I said you bet, so I let our driver go. This was a great start until they asked us to go around the room describing our psychosocial programs. I was really on the spot. I leaned over to Nathalie, who spoke better French and asked if she wanted to do it.

"No, Dr. Kent, I'm too new."

"Just kidding, I'll do it." And I did--not too flowery but adequate. My French teachers over the years, Henri, Arlette, and Cecille, would be proud of me.

Then the fireworks began. The NGOs fell into three categories, Haitian, hybrid national/foreign, and foreign. Several of the Haitian NGOs began to attack and criticize the foreign NGOs for their spotty, inconsistent inadequate care of the Haitian population—how they were overlooking and neglecting large sections of the region. Moreover, they felt that this was being done without proper consultation with the Haitian NGOs and government, who knew where all the underserved groups were, and how to set priorities. They felt that no sectors of the population should be excluded. Though couched in the right terms, the accusations were hot and stinging. Serge responded articulately, saying that all the aid being given, and the NGO support structure to provide it, had been created quickly and efficiently, considering the circumstances, and while accomplishing this first phase the foreign NGOs never had intended to serve everyone all at once. They just didn't have the resources to do it. But now that they were up and running, they would welcome any help they could get to begin seeking out particularly needy underserved groups, and setting new refined priorities for accomplishing this.

Though Serge spoke these thoughtful welcoming words, the Haitian groups apparently needed to let off more steam, repeating themselves, some more eloquently, some just plain getting angry and acquisitive. But finally the hybrid Haitian-foreign groups began to play their swing role, and by the end of the meeting a more harmonious rapprochement began to emerge, both sides meeting informally afterwards in shifting small groups to make networking connections and set up working meetings. Everyone wanted a piece of the action, the power, and the prestige.

Nathalie's eyes got big in the heat of the battle, and I slid down in my chair, glad to be out of their gun sights. Several people dropped by afterwards, saying they liked what I had said, that, instead of just doing direct, take-over work, we were seriously engaged in collaborating and teaching Haitian doctors and nurses to improve their own mental health diagnostic, treatment and triage skills. They realized we were aiming at making a durable sustainable contribution to Haitian career building and their practice of psychosocial medicine. The old adage, "Giving a man a fish feeds him for a day. Teaching him how to fish feeds him for a lifetime," was mentioned more than once. We weren't doing big volume mental health care, but we were giving skills to help their professional practice over their career.

At the same time, I didn't underestimate the heat of the political issues simmering just below the surface. I wasn't naively lulled by the seeming rapprochement at the end of the meeting. The Haitian NGOs, like the government, were feeling invaded and by-passed, not simply on an idle polemical nationalistic basis, while the foreign NGOs were feeling abused and misunderstood. They, of course, knew this familiar process, this sequential dance, and yet it hurt and rankled, nevertheless.

I had been so immersed and consumed getting my own programs up and running that I hadn't taken a look at the big picture. I suddenly had a chilling, overwhelming thought. What I was doing was just a drop in the bucket for a large country flooded with urgent psychosocial needs. I felt diminished and depressed, but was able to bounce back after a few minutes. I no longer found the Cluster Meetings boring. It was a real wakeup call. And yet in no way would I downplay the value of our effort. We just needed to sustain, increase and distribute our training more fully, by engaging in better coordination, collaboration and prioritization. I was reminded of a Mother Teresa quote my patient I tried to Skype earlier in the month had placed at the bottom of her email, "In this life we cannot do great things. We can only do small things with great care."

During the meeting I heard about an eerie thing. One Haitian NGO talked about foreign NGOs being naïve and needing street-wise Haitians to help them navigate through treacherous confusing tent camp waters. He used 'ghost tents' as his example. *What in god's name are 'ghost tents?* I wondered. In order to qualify in tent camps for precious daily food and water rations, you had to have a tent IN the camp. And the more tents you had, the more rations you got. Now tents are in short supply from legit NGO sources, and cost a huge amount on the black market. But on the other hand, qualifying tents don't have to be too fancy. Blanket- and tarp-covered lean-to's count. And food and water remain precious commodities, especially in certain areas. So a few enterprising Haitians hit on the idea of, you guessed it, Medicaid Fraud, oops, I mean 'tent fraud'. On the other hand, someone told me later such tents were the only entrée for whole destitute extended families living outside needing food and water—kind of like what our cook Crystal might be doing. Nothing is simple in Haiti.

A picture began to develop in my mind. I began to see these huge sprawling tent cities as living organisms, not just static encampments. I knew that property records in Haiti weren't very good in the first place, and much of what there was had been destroyed in the earthquake. So figuring out the property ownership jig-saw puzzle would be daunting, time-consuming, and move at a snail's pace.

And the collapsing urban buildings had squeezed the lucky living out into these myriad tent camps, leaving no habitable dwellings behind, and no other place to go if country cousins couldn't take them in. There was talk of creating new earthquake resistant settlements as well as repairing existing dwellings so people could return to Petit Goave, Leogane, or Port-au-Prince. But since most of the displaced Haitians had no work or money, and in tent cities they were getting shelter, water, food, community and safety, as well as on-site (or nearby) health care, these tent cities were currently the best ticket to stability and security. I was beginning to realize that a big problem was developing with the tent cities, creating a whole new urban demographic community of a fairly permanent variety, without much hope for change on the horizon. The tent cities and their denizens would be around for a long time, I was afraid. It was a rare moment of revelation for me, having a chance to put up my periscope and survey the new Haitian landscape before submerging myself once again into the troubled waters of my last few days in Haiti.

As the meeting wound down, and started to run over, I began to look at my watch. My Notre Dame patients would be waiting. I was getting restless. Finally it broke up. But then the informal clustering began, and Serge was in the thick of it. I began pacing, and then whispered in his colleague's ear. He gave Serge the high sign. He broke loose, and we headed for their car. Well, not a car exactly. It was painted white with bright red stripes, a Canadian Red Cross ambulance. We piled in and took off. Traffic didn't get in our way. The potholes did. I was holding onto my bright red backpack. It fit right in with the ambulance décor. Nathalie was bouncing along beside me, clutching her notepad.

We got there with little time to spare. For once everyone was on time. And, both patients were in better shape. So we felt good about proposing the idea of seeing them again in two days, not one, on Wednesday, which would be my last Notre Dame Clinic. I again clarified the plan for my medical transition, and Nathalie doled out the medicine. Compared to the crazy antics at the Cluster Meeting, our patients were very civilized and mature.

As we walked out to the hospital gate, Nathalie ran into a nursing school friend of hers. I had just found out it would be almost a half hour before we got picked up, so she went off to have coffee and I sat down to collect my thoughts and make my meeting notes for Stephanie. Ideas began flooding into my mind, based on the meeting, other observations I had already made, and conversations with other NGOs over a few Barbancourts at the Royal. As I recorded my notes on the meeting, I really got revved up, my first chance to begin putting some things into perspective. Here's what emerged:

I began wondering what brought so many NGO groups to places like Haiti, and how they staffed their Herculean efforts. Did only 'disaster junkies', as Jim Srodes, my journalist friend called them, thrive on it. Or, were there a range of professional motivations, as suggested during that breakfast discussion in the Petionville Residence? As I chewed on this, the idea of 'disaster junkie" began to intrigue me. They seemed to be a special breed within the echelons of disaster relief workers—a distinct but elite few. These were the folks who jump from disaster to disaster with their NGOs, fueled by the high-octane adrenaline rush of each new catastrophe. They even hopped from NGO to NGO, if necessary, to find a new stable, any stable, to feed their addiction. Often consummately good at what they did, they never seemed depressed or discouraged, buoyed and self-justified by the intensity of the situation, their favorite anti-depressant, feeding on the tragedy of the displaced and downtrodden. Pardon me for sounding cynical, but I was just a naïve inexperienced newcomer, still reeling from horror and sadness, and not sufficiently professionally inured to all this suffering, using only my flimsy defense of gallows humor. These 'disaster junkies' all seemed to know each other, and were familiar with each other's organizations, inside and out. They didn't need flow charts, already knowing in their heads where they stood in the NGO pecking order. They spoke of first tier and second tier NGOs without skipping a beat. If one disaster petered out, or their funding folded, they were nimble at organization-hopping, showing up to join their buddies at the next dramatic scene, raring to go.

Many more thoughts were percolating in my brain as I sat there. Luckily I got a few down on paper before Nathalie was finished with her coffee and our driver arrived. I would have to wait until another time to pull the rest of them together.

Back at the Residence that evening, I saw Dr. Paul hunched over a little piece of paper. "You look like a medical student preparing for an exam, trying to reduce everything you know to one small piece of paper. I had a classmate like that. He was a terrific guy, happened to be a Jesuit priest. He once got me to order Rocky Mountain Oysters so he could taste them. I wondered if I would have to spoon feed him since he acted like he couldn't touch them. Against his religion I guess."

He received this with a big smile, complaining, "Kent, you're distracting me," barely looking up from his calculations. "It's the only scrap I could find." He continued working like a beaver.

"Whatcha up to?"

"My statistics, seeing what kind of kid and teen cases I've been running into. They seem to be shunting all the kids to me, being a kiddie doc. Here, let me show you my percents: Colds 20; worms 12; Pneumonia 9; Malaria 7; Tinea (skin fungus infection) 8; Diarrhea 6; Scabies 6; impetigo (boils) 5; psychiatric 4; failure to thrive 4; pain and arthritis 3; Anemia 3; Vaginosis 2; eye problems 2; burns 2; trauma 2; chicken pox and measles 4; and normal kids 2."

"And remember cute little Cassandra, with her pretty dresses and fashion show? She upped that number, counting her worms only once."

"What's all this vaginal infection stuff?" I said. "And in kids?"

"Well, I did have one raped 4-year-old, torn up and infected down there. Made me weep. The Haitian nurses had to help me. They're compassionate, but tough. They've seen it all. They're really disgusted with the few bad actors out there, but assure me most Haitian men, and especially Haitian dads, love their little girls. The other vaginal stuff, just poor hygiene, or uncertainty about normal discharge with periods."

Always upbeat, Dr. Paul threw in, "Obama came by to see me the other day. I was really excited to see him. He actually gave me a high-five!"

"Oh, come on, Paul, what are you talking about?"

"This 1-1/2 year old was born the day Obama got elected, and that's what his parents named him. Oh, and there was actually a John Kerry, too. I was glad there was no John Edwards."

"Okay," I said, "Those were the high points, and the 4-year-old girl the low. Anything else notable?"

"You know those cute little goats, and that baby pig that came in the tent to visit me the other day?"

"Yeh?"

"Well, the mother pig came in looking for her baby, must not have trusted me and took a big poop right on my left shoe."

"Gross. So?"

"You're not plagued by all the things I worry about. The cysts of the pork tape worm, might be in that stuff; I get a little on my shoe; at night I take my shoes off, get a little on my hand; I brush my teeth, a little gets in my mouth. And Voila, Dr. Paul's got his own pork tape worm!"

"You gotta be kidding, Paul? I think you've been listening to Alisia too much"

"Nope, it comes naturally."

Later, Alisia took me aside, "I'm a little worried about Nathalie. She keeps looking strangely at Dr. Paul.

"Well, just ask her," I said. "She speaks good English."

Later she told me what Nathalie said. "Dr. Paul showed me a picture of all his kids. You know, his wife is white, and most of them are different shades of brown. I never would have thought Dr. Paul was THAT kind of man, all those different mothers."

"You should have seen Nathalie's face when I broke the news they were all adopted, and from Haiti, I think.

She said, "Oh! I feel so much better. Good for Dr. Paul!"

I curled up with my computer, with lots to capture, not realizing what was in store for me the next day.

Tuesday, March 30, Miragoane: Shocks

I awoke with a start at 5:30 AM, and looked around. Tom was barely stirring. Time for exercises. I walked out in my shorts, surprised to find Alisia and Paul.

"Did you feel it?" said Alisia. "It was like the subway blowing through under my apartment."

Both had felt it, but I had missed another aftershock, awakening because of it. Damn! But her description immediately told me what it was like. I had been fishing on Friend's Creek near Gettysburg one time, guest of Judge Tom Hogan. I found a perfect pool, a hatch was just beginning and the trout were rising to strike. I was just beginning my cast when I felt a rumbling tremor under the stream, transmitted right up my feet. *Oh my god, an earthquake!* I thought. But nothing else happened. Being a doc, it crossed my mind maybe I had had a little seizure. My line lay tangled downstream behind me. I began to reel in, preparing to cast again. Just as I was about to, the same thing happened, and this time I distinctly heard a deep subway rumble along with the tremor. My diagnosis shifted. I was clearly hallucinating, and it scared me. After a third episode, embarrassingly bizarre and distinct, I packed up my rod. It took me quite a while to dare mentioning any of this at dinner. Tom laughed. "You're not hallucinating. There IS a subway running under the stream at that point, way out here in nowhere."

"What the hell are you talking about, Tom?"

"And you paid for it. It's our high-security federal government command center in case of nuclear attack. It's under that mountain and honeycombs throughout this whole area, including a subway system connecting it all."

So, yes, I knew what Alisia was talking about. Only no actual subway in Haiti. Something much worse.

With this start to the day unsettling me, we headed out to Miragoane for my last clinic there. I was expecting our dingy, low-walled back room, but Dr. Affricot, our Haitian doctor *de jour* and Chief of Clinic, offered us his office. I teased him about his President Preval impersonation at the Saturday Seminar, and we got ready for our patients. We took care of three follow-ups, our man with tongue biting who was having many fewer attacks, our psychotically depressed 51-year-old on Amitriptylene, who was eating and sleeping better, and our 8-year-old agitated seizure who had calmed down somewhat and was seizure-free so far.

We then worked our way through several patients, a moderately depressed 74-year-old woman needing medication and social support, a 19-year-old boy with learning difficulties and mental retardation, who spoke to himself all the time irking his parents and keeping them up all night. He needed an anti-tension sleep med as a first step.

The second to last patient was a pleasant boy, originally a bright capable student, until sustaining a head injury, resulting in a depressing decrease in his abilities, making him now just average, which frustrated him and confused his parents. He alone seemed to understand how much he had lost. We explained the narcissistic injury that resulted in his irritable mood, and pinpointed an additional, post-concussive source of headaches and attentional difficulties. Explaining this to him and his parents helped a little. We then suggested, if they had access through their pharmacy, they might try methylphenidate (Ritalin), which could help his concentration. The last case was an anxiety and stress problem amenable to support, imagery, and re-breathing techniques.

Wrapping up, I carried chairs out to the waiting Patrol car, feeling sad it was all over, the last Mirogane Clinic now behind me. I was sitting alone, lost in thought, feeling a little strange. I had to wait quite a while for the others to wind up and come out. The driver was buying a cola, which I couldn't afford at the moment. I happened to look in the rearview mirror and saw a rather tall, dignified, well-dressed man walking along the road toward the van. He was so formally attired I found myself thinking of Papa Legba, the seminal voodoo god of the crossroads, the one standing between the underworld and afterlife, life and death, and the past and the future. The man had a cane like Legba sometimes does. I began to get uncomfortable, sinking in my seat. I felt somewhere between a dream and a nightmare. When he got parallel to me, he pivoted, bowed his head down at my window and rapped at me with his cane. I debated whether it was safe to roll it down, fearing his fateful words, but finally did so.

"Pardon me," he said in formal French. "Are you a doctor? I see the IMC doctor sign on the car."

"Yes."

"I have a problem maybe you can help me with. Do you treat Hensen's Disease?"

I said, "Well, I don't know what that is. You see, I'm a psychiatrist."

With that, he tipped his hat, swung around, and walked slowly off down the road. I felt like I had failed him, flunked my last test in Haiti, and missed my chance to help, though he seemed to carry on somehow by himself. The driver had heard the last part, came over, and explained that Hensen's Disease was the rich man's way of saying he had Leprosy. I was startled. He looked perfectly normal and healthy, yet his flesh would eventually melt away if untreated. Why hadn't I been able to do more?

Quickly I looked out the window to catch sight of him again, but saw a dog up ahead instead, one of those quintessentially Haitian breeds. It was walking along the side of the road up where the man should have been. What had happened to the man? Everything was starting to feel surrealistic. Suddenly the dog turned to cross the road. A truck flashed past my window, hitting the dog broadside with an awful thud, spinning the creature round, running over his right hind leg. Yelping shrilly, then baying mournfully, the dog barely dragged itself to the side of the road, fell over twice, then somehow righted himself on three legs, hobbling down the road, dragging his crushed right leg, eventually crossing over and out of sight. I don't know why, but these two experiences fused in my mind, putting me in an awful frame of mind. I felt sick to my stomach, then lapsed into a deep sadness for the chronic illness eating away at a man's flesh, and the dog's near mortal injury. Yet the man walked on and the dog picked itself up again, and hobbled off. *Were they one in the same?* I wondered. *Had my mind, or Legba, played a trick on me?* I could feel a struggle going on deep within me, my dreams for Haiti being threatened by the return of a nightmare again.

When the others, chattering amicably, climbed into the car I was strangely quiet for quite a while, suffering flashbacks. I was approaching the end of my journey in Haiti and was deeply worried about her future. My mood was somber, as crumpled houses and tent cities floated by like a macabre dream. The earthquake had ravaged an already crippled country. Had I made any difference? I had come down with such hope and idealism, wanting to help my friends and their country. The previous Monday's Cluster meeting came back to mind. There was still so much to do for Haiti. I began to feel very small and insignificant.

Only the word 'penis' roused me from my reverie. Did I really hear that? I tuned in quickly and was surprised at the animated conversation going on around me. I had thought Haitians were quite Catholic and a bit prudish, but to my surprise one of most open and explicit sexual conversations I had ever heard was steaming around me—maybe because they were a bunch of medical professionals. A nurse had just said young men only think with their penises and never listen to a woman (*zozo rad sans oreille* [a hard penis has no ears]). One of the doctors countered, saying that that was an unfair male stereotype; he for one didn't like sex unless it was with someone he cared about. The other nurse felt men were only after their own pleasure, not taking time to think about what a woman needs. The other doctor disagreed, saying a man really wants to please his woman, but women were often shy or afraid to say what they really wanted. Nathalie spoke up, saying it was true, often hard to be frank with a man, because he had so much wrapped up in pleasing and performance. She went on to observe that, unfortunately, there are some women who have trouble with orgasm and fake it until they really know a man and can talk, if they dare. Some women like it hard and fast, fitting in easily with the old male stereotype, while others, like her, only like that occasionally, and prefer to have it slow and tantalizing, something that can be exciting but takes time and timing. And not all men can hold on. The first doctor said he knew what she meant, at least for the first go round in an evening. But the second time was a different story.

Soon they were all agreeing that a more enduring intimate relationship allowed for more pleasure and perfection. A nurse suddenly laughed giddily and blurted out, "But a hot, no-strings-attached, one-night-stand can be incredibly exciting too, if you're careful about precautions and STDs (sexually transmitted diseases). Variety is the spice of life. I remember the days, before I was married." Everyone laughed.

"Listen to us," one of the doctors quipped. " You sound like a man. We've come full circle."

"We're modern women!" she said, "Or, I used to be."

Nathalie then summed up, "But sex in a loving relationship is the best, and it's enduring." Everyone agreed. Sitting there amazed, I privately added my own agreement. *Love in the time of Cholera* floated through my mind. Yet I found myself feeling very encouraged. Life, love and baby making were in the air. I was relieved and hopeful for Haiti once again.

Notre Dame Clinic wasn't until end of day tomorrow, so I had a little time on my hands. The driver deposited us at the Residence, and Nathalie and I settled into the living room easy chairs. "Tomorrow is the Boat Clinic, Nathalie. You're in for a treat. It's a beautiful boat ride and a fascinating place."

I saw Nathalie shudder and freeze. "I can't do it. I can't get near the water or go on boats," she said in a whisper.

I knew instantly what was wrong. "Your father, right?"

"I haven't been able to go near the water or swim since."

"We need to talk," I said. And talk we did, for over an hour, from every angle. I finally said, "It's an essential part of your job, one you have to find a way to do. This may be one of the toughest challenges you'll ever face, but professionally, it is a real opportunity to overcome something that your father would want you to do. And you'll be helping patients do this kind of thing the rest of your life. If you don't, and harbor this sad secret in the back of your mind, it will be hard to have honest conviction in your work."

She remained silent, looking a me. "You did love the beach and swimming with him before, didn't you?"

"Until 10."

"And you were good at it?"

"Yes."

"Then you can be good at it again. But starting tomorrow, we'll put you between the two boat nurses, in the middle of the boat, and hope for a nice calm day."

"Do I have to?"

"Yes, but with me and Tessier and two other doctors and nurses. You'll be safe in reality, and will begin to overcome this so you can feel safe inside. You don't have to like it, just tolerate it." She didn't say anything. But she didn't say no. She was thinking.

There was another cliffhanger going on, and I was down to the wire. Tomorrow I was going to see Eustache for the last time, the tentless Boat Clinic nurse, and I had virtually promised her I would get her a tent. And I deeply wanted to. But I had no money, and there were none to be had anyway. But there was a sad drama unfolding that I knew about, with the chance I might be the beneficiary--if I were lucky. My namesake, Dr. Paul, was in the center of it. As a matter of fact, he was sleeping in it. I knew he wanted to give his tent away before he left, and tonight, Tuesday night, was his last night with us. There was only one problem. Stephanie had claimed the tent ahead of me for a compassionate purpose. You see, there was an infant who had been dropped off secretly in front of a friend's house. She heard the infant crying, picked it up, reached for her phone, and called Stephanie. Somehow, through networking, they were able to identify the mother, who was in a bad way. She had no tent, no food, no water, nothing. With help, she was possibly coming around and might take the baby. And Stephanie wanted her to have the tent if she became competent enough to take her baby. The timing was nip and tuck.

We heard the brakes of the Mongoose, and Stephanie stepped in. I hardly dared ask her, but eventually I had to.

"Oh, the tent?" she said. "We won't be needing that. It seems the father found out about all this, and has claimed the baby and the mother has disappeared again. He was neglectful, but he is competent, and he has rights. And we all think the baby will be in good enough hands. Why do you ask?"

"I don't know if you remember, but the nurse in the Boat Clinic doesn't have any shelter."

"Oh, right. By all means, give the tent to her. We should do everything we can to shelter our own."

I looked over at Dr. Paul and said, "Are you okay with this?"

"Absolutely."

"And, Paul, I'll make sure that Eustache knows that this is the "Dr. Paul Kent Memorial Tent", from both of us."

He laughed, "That's terrific. My wife and I both love Platon. That's a great use for my tent." Relieved, I also gathered a spare mosquito netting and a poncho to go with my air mattress and camping pillows, planning to divvy this up with Marie, the other nurse, who might be a little jealous. I packed all this away for the morning. We needed to be organized. We would have our work cut out for us

Wednesday, March 31, Platon: Challenges

The next morning was cloudy. It had rained during the night, waking me up. Apparently rain gets to me but not earthquakes—maybe because any downpour could be the beginning of the rainy season. I was worried about Nathalie and Platon. It would be very sad to miss going there my one last time. I wanted the weather to be nice for her sake. After my exercises, I went out on the portico and looked to the east. Slivers of sunlight crept out through breaks in the clouds—sending rays of hope. It was clearing up. We were in luck.

At the Office, I took Eustache aside, and told her about the tent and things. Her eyes lit up. I showed her where I had stashed the tent, then explained I had put some other things there too for Marie, so she wouldn't feel left out. With a huge smile, she called a friend with a car and said he would pick it up on the way home. She gave me a big hug, but I put my finger to my lips. I then explained Nathalie's situation and arranged a supportive loving sandwich for her between the two nurses. I urged them to chat it up, distracting and keeping her focused elsewhere. When I came out Nathalie herself was already chatting with one of the drivers, dressed in a chic, colorful outfit, a bit low cut again. Tessier cruised in a little late, we grabbed a few chairs, and were off. I said some encouraging words to Nathalie along the way.

When we got down to the boat, Nathalie was hanging back, clearly not dressed for wading out. To my surprise one of the gallant boat guys scooped her up, carried her out, and kerplunked her right in the boat between the two waiting nurses--teamwork I'm sure they arranged to everyone's satisfaction. She didn't say a word. But her eyes were wide. The nurses put a stop to that by actively engaging her in conversation, chattering away as prescribed. Pretty soon, Nathalie was chattering back at them. I was really proud of her. She was a lot tougher than she looked. We were off and running. I could go off duty a little, taking time out to enjoy the scenery and surf through my own thoughts.

Looking at the mauve mountains, with their crinkled valleys besequined occasionally by patches of bright green, I found tears coming to my eyes. This was goodbye, my Haiti Farewell, and also a fervent wish for Haiti to Fare Well. I had been so eager, yet so anxious and unsure, when I landed on her shores again. But I was now satisfied that I had given her my all, and had passed my biggest final exam.

If I had sailed off with that stock broker and his goddess on the Mektub, leaving my future to fate, I might never have gone to medical school nor developed the skills allowing me to help Haiti in her time of earthquake need. It was a fateful decision. After a near miss, I stayed in Haiti and ground out my fieldwork on a shoestring, poor in cash but richer in knowledge and spirit. Occasionally I wonder what it would have been like on the Mektub. Where did they ended up? My depression lifted slowly that second summer long ago, followed by some good solid ethnographic work, and increasing excitement about medical school. I treated a lot of peasants on the fly, but knew I didn't really know what I was doing. They were grateful for my efforts, my neophyte medical help and caring. And they repaid me with friendship and support. I was a very lucky guy. I returned to Boston in time to buy my books and go on to medical school.

So here I was fifty years later, motoring along Petit Goave Bay, staring out at the Il de la Gonave, heading for my last visit to Platon, glad of my long ago decision, feeling sad again about Haiti, surrounded by my wonderful new friends, and very much looking forward to seeing our patients.

The boat trip there was not the usual one. First of all, we went past two boats doing seine fishing with a huge net. They were gathering in silver flashes glinting in the sun hinting at an ample harvest as we circled. But not for long. One of the doctors and our captain noticed dugouts they recognized, and pulled sharply in toward them. Soon the fishermen were holding huge lobsters up by their feelers, tails flapping and snapping in the breeze. They were prime creatures. A typical Haitian bargaining session began, soon turning into a seemingly cantankerous fight. We pulled away suddenly, leaving everyone empty handed—only to be called back by a lower price. Finally all the lobsters came flapping aboard, to be divvied up later.

One thing I knew. I would be eating one of them that evening at the Royal, by hook or by crook.

We arrived at Platon and I intended to set up shop under the Tamarind tree, only to find fishermen's nets filling the whole area. Soon Eustache had a little girl pulling them away from our office, using a makeshift straw broom to clear our spot. We had hardly taken our seats when Alice St. Leger came striding in, not quite so high and buffoonish, talking more coherently, and thanking us profusely for the tent and the food. We got her to sit down, and took her pressure. It was still high, but a more healthy 160/90 and she hadn't stroked out from the lowering. We introduced her to Nathalie, who chatted with her in Creole, as we set about lowering her pressure still more. Everything was heading in the right direction, but whether she needed a mood stabilizer, or a miniscule safe amount of Chlorpromazine, was not yet clear. Nick and Dr. Beauge would have to make that judgment call. Small amounts of neuroleptics might be okay, though even moderate amounts can add to stroke risks in someone her age in her condition. All that would come later.

We saw several patients our last day, including men, women and children with seizures disorders. They were coming out of the woodwork. I was becoming a real expert.

But there was one case of a different sort that stood out that final morning. This 71-year-old gentleman with a tattered vest sat down, and had all the hallmarks of a fairly severe depression. But his condition had gone on some seven years. He complained of poor appetite and significant weight loss, and, indeed, he was emaciated, ribs and breastbone sticking out. He also had urinary frequency and difficulty initiating his stream. *Was this from depression-related constipation pressing on and constricting his waterworks, or maybe benign prostatic enlargement? Or could it possibly be prostate cancer? Too slow a course,* I thought. I took a look at the clinic slip to check his pressure. It was surprising low, as was his heart rate. Dr. Beauge had checked it already, 52. When I took his hand in mine, it was very cold. His basal metabolic rate had to be very low too, then. So we both wondered if he had low thyroid, even a touch of Myxedema madness, a change in mental functioning sometimes accompanying very low thyroid. But the symptoms of depression were so striking that we opted for giving him both Amitriptyline and a modest starting dose of thyroid. What an interesting last case, one stretching our capacities for dual diagnosis and giving us a chance to help this affable, slow-moving guy.

We were packing up shop when Eustache came running over in quite a flap, a late patient in tow. Both were looking guilty.

"Here she is, Dr. Kent."

"What do you mean, Eustache?"

"You know, that movie star--her patient."

Oh, my god, I thought, *I can't believe I totally forgot.* I reached into my backpack, fished out my wallet, and found the note from Sienna.

"What's her name, Eustace?"

"Lexina."

"That's her."

Actually I wasn't sure that's what the note said. It had gotten wet and smudged on the boat ride. So I had to trust Eustache.

Lexina, 33, was petite, thin, and drawn, complaining she couldn't think or remember things, had no appetite, and kept having an awful smell in her nose. She had obviously lost weight. She was also having panic attacks. On further questioning, she complained her heart beat fast in spurts and she had shortness of breath. Her nights were sleepless.

The smell made me think of a seizure aura, maybe because I had just had a spate of seizure patients. But the rest of the story didn't fit. I recalled Sienna feeling Lexina was quite depressed, but I was seeing more anxiety in her now.

"What happened to you?" Dr. Beauge asked.

She told us she had been down by the seaside with one of her children when the earthquake hit. She became paralyzed with fear as she heard the screaming, not able to turn around and look. Soon there was a terrible smell in the air. Finally she turned. Her house had collapsed. One of her children was in it. The meat she had been cooking had fallen into the fire, along with a leather pouch, and they were smoldering, producing a terrible stench. Some of the thatch had caught fire too, threatening to burn the house down.

Her thinking jammed and her mind went blank. Finally she forced herself to go up and look. Her eyes searched desperately. Suddenly she heard a noise, and saw her child run out from behind a tree, crying hysterically. She was safe. But the psychological damage had been done. A panic reaction had been detonated. Now two plus months later, her acute traumatic reaction had been moving toward a more depressive, post-traumatic phase, putting her at risk for chronicity. She had refused to come see the doctors for some reason, until Sienna and Eustache coaxed her in.

"Dr. Beauge," I said, "Ask her why she didn't come sooner to see us today."

She hesitated, then tearfully told him, "My sister's child actually died in the quake, and she's handling it much better than me."

Eustache pulled me aside, "Dr. Kent, she seems better than last week, not so depressed."

I asked her in Creole if this were true, and she said, "Yes, I'm feeling better, less down, but more anxious. Eustace got that pretty lady talk to me and it helped. She was so kind I could finally tell someone. I was so embarrassed to be that bad off when my child actually lived."

"And guilty your sister's didn't."

"Yes. But I was scared to death and couldn't get the fears out of my head. I kept having nightmares."

I noticed her blood pressure was high, not just her anxiety. Talking to Sienna and Eustache had swung her away from chronic depression back into the healthier zone of doing the scary work of thinking about her experience and overcoming her associated anxiety.

"Why were you so late coming today?" I asked. "I'll bet it was still hard to come, and tell a new doctor."

"Yes, Eustache had to come yelling for me."

"We're all glad you found the courage to come and talk again. It's more than half the battle in working your way back to health. I want Eustache to help you find the nerve to go talk with your sister, about her sad loss and your near one. She may seem like she's over it, but I'll bet not. She needs your help, you know. So you should talk with her."

We made this a therapeutic assignment, with follow-up for the next week, and gave her Atenolol for her high blood pressure, which would also help interrupt her panic cycle and sleep disorder. We also taught her the Sac re-breathing technique for hyperventilation and the Valsalva for rapid heart beats.

She thanked us and started to leave, then turned around with a smile, "Tell the pretty lady I came!" We breathed a collective sigh of relief. My name would have been mud if I had blown Sienna's referral.

Even with all this, it was still early as we packed up. The sky was clear and the Caribbean lay sparkling before us. Finishing early, we had time to burn. I looked again at the ocean, then at Dr. Beauge. We both smiled

"I'm going swimming!"

"Me too!" he said.

What I didn't know was he had a suit, and I didn't. But I had my handy nylon camping briefs on, and so down to the shore we went. Of course, lots of kids and a few grownups followed us. Dr. Beauge quickly changed into his suit, and I simply dropped trow. There were some good-natured snickers, which increased in volume as they saw how tender my feet were picking my way through the rocks to the sand and deeper water. It was warm on the surface, and cooler down below, just like Bananier. Dr. Beauge and I paddled out and splashed around. Kids began swimming out to us in droves, gathering around me in particular. I couldn't resist and splashed the younger boys. They splashed back. Pretty soon we had a major water fight going on. Then I made my two-hand whomping splash behind one boy's head, and they howled.

As we engaged in boyish antics, a bevy of teenage girls joined the fray. One finally asked my name in Creole, then all chimed in, enjoying talking with me. Another girl said, "Will you sing us a song." On a roll by then, I sang, "She'll be coming 'round the Mountain". Then I challenged them to sing, not having a clue I had been set up. These girls, four of them, preceded to sing and do a choreographed set of moves around a Creole pop song. Not to be outdone, I sang "Row, Row, Row Your Boat", paddling away at the water; they replied with another choreographed number, and I finally sang "Old Man River", hitting the low notes, sort of. The aqua-audience had swelled considerably by then. But I was becoming exhausted, and a little self-conscious. I waved them off, and looked around.

I had been so wrapped up in all this fun that I hadn't noticed something that really touched my heart. Nathalie was with Dr. Beauge, out chest deep in the water, with a life preserver on. He was holding her so she floated safely as he taught her how to do the crawl again. I watched with a smile, proud of her but a little wistful. I wished it were me. Talk about a successful, one-day desensitization program. Then the wind started to come up and the boat pilot gave us the high sign, yelling at us to get a move on.

I swam back over to my pants, painfully picking my way back up the rocky shore. Apparently, Nathalie had shed her clothes there, too, and borrowed someone's suit. As I put my shirt on, she began taking off her two-piece top. My fatherly pride shifted gears on me. She was well proportioned. But she was adept at doing the 'bra in the back' French thing, slipping one thing down as the other went up. I lowered my eyes, collected my stuff, and headed to the boat, Nathalie following. No one had to carry Natalie this time. She was already wet and stepped nimbly into the boat.

I was beginning to think of my last Notre Dame Clinic, wondering how it would go, and if we'd get there on time. The only boat we had on this trip was the slow one. We had no volunteer docs this time requiring the second fast one.

We made it to Notre Dame in plenty of time, again going without Tessier. Nathalie and I worked pretty well as a team. I was transitioning both patients to Nathalie as our patient transfer bridge, awaiting Dr. Nick coming aboard. What we found with our 20-year-old patient was that, still seizure-free, he was now also delusion- and hallucination-free, and talking about going back to work. What did he do? Welding. And he maintained he did it without darkened protective welder's glasses. I was somewhere between incredulous and horrified, and trotted off to get him some old exposed x-ray film I knew he could use to protect his eyes. I had been taught this when watching eclipses. Nathalie was interviewing nicely on her own. She could really swim with her patients.

His Haldol was now down to a maintenance amount, and his Carbamazepine at a good level. He was eating, sleeping, and socializing nicely, well on the road to recovery. Seizure-generated manic psychoses can resolve fairly quickly once the underlying seizures are brought back under control. We explained the importance of taking his seizure medication regularly, the transition plan to Dr. Nick, and the idea that he was doing well enough to come back a week later. They were happy the way things were going, and felt coming back in a week was fine.

When he got up, our young patient reached out for Nathalie's hand, holding it for a little too long again, saying, "I'll see you next week. That will be very nice." As soon as she could, Nathalie pulled her hand away. I could tell she was a little shaken. I thought of her pretty face and low cut dress for a split second as this happened, seeing her through his eyes. He had transferred his allegiance to her, but with a romantic tinge. I could tell it really bothered her.

Our agitated woman walked in with a more normal gait, wiggling only a little, and looking me more in the eye. We talked a little of her fears and losses, and then her son-in-law said she had a question.

"Doctor," she said, "this weekend is Easter, and I've been away from my husband for a long time. Do you think it would be all right for me to go home and visit him? We're missing each other."

I was floored, and pleased. "I think that would be a wonderful idea, you two being together for Easter." Nathalie cracked a big smile. Things were moving along. I saw the young man accompanying his grandmother, looking at Nathalie repeatedly. Was it the same as with the other younger patient? I filed it away again.

We packed up shop, stowing the fan and the chairs safely out of site. I was late calling Samedi for our ride back and we had finished our follow-ups early. We loitered out near the hospital gate, until the guys hawking their Haitian primitive paintings drove us a little crazy. So we retreated to some chairs inside the grounds, giving Nathalie and me a few moments to talk. Finally, I felt I had to bring something up.

"Nathalie, you're a great nurse, good with patients and nice to work with. You also happen to be a really pretty girl, with a nice figure, and you dress with lovely stylish care. But I think you noticed how that first male patient held your hand too long, and I think the other guy kept looking at you at the end. You also told me at least one patient stalked you."

She looked at me and said, very slowly, "Yesss, Dr. Kent?"

"I've learned from long experience with people I supervise that when god gives a professional good looks, it is a good idea to 'dress down', rather than in an overly stylish, slightly revealing fashion. The great way you look will already stir up enough attention from guys, and enough jealousy from gals. So you don't need to stir up even more of it. It will just complicate your professional work and your own personal life. This is especially so when you are working out here in the countryside. Because I'm a guy, and not immune to these reactions myself, I think my judgment's pretty good on this, although I could be wrong."

I was going on like this, trying to find words that wouldn't be too hurtful given such a sensitive subject, when I happened to notice that she was clutching her note pad tightly to her chest, covering herself completely. And her head was turned somewhat away from me. Her body language told me I had stuck my foot in my mouth. And yet I felt someone needed to say this to her. You can't be an attractive Vogue model out in the wilderness, especially when you're trying to do psychiatric work. And especially with peasant countrymen.

Well, I had helped her with her boat phobia, but maybe I was striking out with this. Win one, lose one. So be it. I think I was right but I felt like a heel. On the way to the Residence I mused about how awkward I felt talking of young attraction as an old man. I knew, on my side of the fence, I was working on transmuting my aging masculinity into a more paternal giving. We all had things we were working on at our respective stages in life. There was a certain sweet sadness that went with letting go and suffering the growing pains at my late stage in life. It also crossed my mind that being a child psychiatrist might contribute to my being a late bloomer, remaining playful and open to things usually long forgotten.

I was able to borrow money from Stephanie for my last supper, promising to follow her plan for repayment. So I was looking forward to my Royal lobster. Tom and were praying that Strict Badou would be holding court again. This was our Holy Grail for our celebration of my last evening in Petit Goave, the prospect keeping us intrepid relief warriors going from day to day. A Heinekens, a Barbancourt, and Strict Badou. That was our mantra. We all showed up on the Residence portico for the ride over at about the same time. And there was Nathalie, wearing a lovely flowered print dress with about the highest tight collar you could imagine, reminding me of a nun. (None for YOU, Dr. Kent.) She looked lovely and conservative. I had meant my comments about patient work, not social gatherings, and was a bit wistful—but happy to be her surrogate father for a few days. She deserved one.

Thursday, April 1, Petit Goave to Port-au-Prince: Bearing Gifts

Tessier, an accomplished teacher and still out of a job, asked me to give his resume to Stephanie. She knew I held him in high esteem, and said she would consider him for positions coming up. As a highly deserved parting gift, I gave Tessier all the money I could spare ($20 sadly), a brand new portable CD player, and Norbert's prized Swiss Army Knife, saying, "Tessier, it's been great. You're a friend for life. And you know what, I wish I could give you a psychiatric diploma. You're better at asking patients the right questions than many of the doctors. You weren't supposed to butt in so often, but you went to so many teaching sessions you couldn't help yourself at times. I had the same problem, as you know. And you were great." We gave each other big hugs, and said goodbye.

While all this was going on, out of the corner of my eye I saw Stephanie lead Crystal the cook into her office. Nathalie and I kept working away at the final pharmacy psychiatric medicine count, making ready to hand things over to Dr. Nick. A while later, I saw Crystal come walking out, moving slowly, eyes moist and down cast. I had the fantasy she had been fired. But knowing how many things Stephanie had on her plate, I suspected she had just chastised her, set the rules, and gave her one more chance. Her family army might go a little hungry, but she would still be bringing home a salary, given the 90% post-quake unemployment rate in Haiti. I hoped so. In any event, the dastardly deed had been done—by someone else. Because Nathalie had to rush off to the departing Beatrice Clinic Patrol Car, we somehow missed saying a personal goodbye, which saddened me, and yet I understood—maybe too much. I gave Stephanie a big hug goodbye. She was particularly warm. Next came Jattu. We hugged with big smiles. Joanne and I gave each other hearty goodbyes. Out front, with my driver waiting, I said goodbye to the rest of the guys, and especially to my man, Samedi. The only thing I regretted was not having my camera to take everyone's picture. Well, they would all be held clearly in my mind's eye, and close to my heart.

When I went back to the Residence to get my suitcase for Port-au-Prince, I surprised Crystal. She was sitting alone in a chair looking subdued. I gave her some parting gifts, including slightly worn running shoes "for some friend of yours." She actually smiled. I had the feeling I would have gotten my scrambled eggs the next day, and on time. Ruth, the laundress, got a few shirts, in addition to whatever else was already missing. And off we went to Port-au-Prince.

The drive back was nostalgic, sad and disturbing once again as we passed by Leogane and Brache. I talked the driver into going slowly through my old drop off spot near my field site. For the first time I was able to pick out where I had gotten off the Camionettes in the days of yore, sorting out familiar landmarks from newer buildings, all mostly collapsed. As we drove, I saw all the HASCO sugar cane fields, and the old molasses mill. Memories came flooding back, along with fresh tears for the destruction I saw all around. There were signs of clean up and some rebuilding, though still precious little. It would take a long time, vast amounts of money, incredible effort, and cutting lots of red tape. And I knew I wouldn't be part of it any longer, at least directly.

A fresh wave of sadness hit me with those thoughts. And yet I was also somehow smiling through it all, a sense of having survived my challenge, my final exam, eventually with pretty good marks. I had made my contribution. Once again I had an abiding sense that Haiti had given me as much as I had given her. Though buildings had been flattened, and the social fabric rent asunder, what I loved about Haiti was her fabulous people and their remarkable spirit. That was still intact, at least among the living. And for those alive whom we met who had fallen below the safety net of their formidable emotional resilience, my colleagues and I had helped provide substantial assistance.

I know I am waxing a little idealistic, like the last time I left Haiti. So I need to remind myself not to gloss over things. The earthquake was horrendous, the impact of the devastation ongoing, and the challenges ahead almost insurmountable. And as Tom said, the hardest part of all this is finding a way to help the Haitian government and the NGOs get their act together.

All this is real and daunting. But I must admit that, in addition to struggling with these exterior things, I couldn't avoid my personal interior battles. Revisiting Haiti under duress stirred up angels and demons lurking in my private abyss.

As I flew down to Haiti this time, with my tectonic plates sliding, I wondered, as you know, if this unexpected return would be a dream or a nightmare. The first hints I was becoming unglued took the form of reminiscences good and bad, and that black tarantula crawling onto my shoulder, followed by fears of infection and emotional dis-ease. My thinking seemed a bit loose and crazy during those early nights, my concerns bordering on the paranoid. My sense of competence and masculinity went partly out the window, my need for survival and personal care coming to the fore. My tent 'madness' and urgency to write my diary for decompression gave some measure of my distress. All of this was fueled by the privilege of hearing the intimate stories of our patients, taking me close to the 'belly of the beast', and bringing me perilously near my own vulnerability, given my aging body and senior mind.

By being open to all this, I became one with my Haitian patients and doctors, using my loss of identity as the wellspring for empathy and understanding of their struggle. Getting in touch this way served as my best guide to helping them. It provided the alchemy for all the things we were able to say to our patients. Our words gave them a sense of being truly heard and understood. The intimate accuracy of our comments became our most potent medicine, delivered with loving care and deep conviction. We wanted to instill hope and a path toward health, allowing them to go on living even in the midst of turmoil and despair. I allowed myself to be possessed by shared demons, revisiting my nightmares to know theirs. I discovered I had left a lot of unfinished business in Haiti, and the gods and the earthquake gave me a chance to come to grips with them all again, but from a new perspective. After reading my diary, my dear friend Maureen wisely observed, "Haiti really has provided amazing bookends for your life."

I was surprised when we entered Port-au-Prince, though I probably shouldn't have been. It was more than three weeks later, and not much had changed: rubble piled high, refuse everywhere, unsavory smells of sewage mixed with delicious wiffs of street vendor pates, with swarms of people and buzzing chaos all around.

After a lively debriefing with Lynne, Peter, and Nick at the Plaza Hotel, I took Nick aside and asked him to take pictures of the doctors and nurses at his next Saturday seminar, since my camera was broken. He was kind enough to send them to me. I conveyed the message that everyone wanted certificates after our course. The doctors and nurses were quite pleased with their seminars, participated with interest and enthusiasm, and felt they had put in a lot of effort and learned a great deal. We had given a pre-test at the beginning of the seminar series, and a post test after each. The medical and nursing groups were rightfully proud of their accomplishments.

IMC Drs (up L-R) George, Philogen, Affricot, Louis, Beauge, Judson; (lower, L-R) Guirlande, Louisjis (Dr. Polo absent)

IMC Mobile Medical Clinic nurses. Nathalie second from upper L. Marie middle L. Eutache lower middle

I was driven back to the Port-au-Prince Residence, where I had my last night's sleep, fittingly in a tent. I had given away my air mattress to the other Platon nurse, so I was a bit apprehensive, until I saw the tent had a cot—a nice upgrade.

I had promised on bended knee to repay Stephanie when I arrived and received my per diem in Port-au-Prince. I was to return it to her friend, Hannah, the IMC Human Relations Director. Scurrying around, I finally found her. Her smile, even as she took my money, could melt a man's heart. My other mission was a self-appointed humanitarian one. Nearly blind from lack of night lighting at our Petit Goave Residence, I marched over to the Logistics Office, and after much pleading, and an eye-test, I convinced a helpful logistics guy to requisition 10 50-watt fluorescent power-saver light bulbs to be ferried out to my equally blind buddies at the Petit Goave Residence. I called ahead to let Tom and Jattu know they were on the way, so they could keep a look out for them. I didn't want them to go on sale on the street outside our Petit Goave Residence.

Things always seem to happen to me while I'm sitting on the couch with my computer in the Port-au-Prince Residence. In walks Crystal again, this time with a bright, serious low-key woman, Nancy Aossey, who happened to be the President and CFO of the International Medical Corps. Crystal introduced us and Nancy asked how things had gone in Petit Goave. Her time was precious, so I gave her a quick overview. She asked a few penetrating questions. Then I added, "You run a great organization, very supportive, and doing terrific work. This has been an experience of a lifetime."

"From what I hear, you were much appreciated out there."

Being in the Corps, I felt like saluting, but instead I said, "Glad I could help." She smiled and moved on. What a fitting exit interview, meeting the boss.

I had half-expected a sleepless night, with howling dogs and crowing roosters, and was almost disappointed when I awakened to bright morning sunlight. I had slept through the night. I guess I was more relaxed. My unconscious was quieting down. That last morning, Pierre cooked me an omelet. Actually, I had to explain how to make it. He thought I wanted him to put it all raw in the blender, and whirl it into a creamy froth, like some health food nut. He was relieved I wanted him to cook it. It was delicious. Maybe that's what had stopped cook Crystal, though I doubted it. I made Pierre the beneficiary of Norbert's caterpillar mosquito contraption.

There waiting at the door the next morning was Bennet, my transportation 'bookend' driver, the same guy who had picked me up at the airport a month ago. He was the airport specialist. We talked of our experiences as we drove an interesting new back route. I realized it was the route Nick and I had been looking down on from our perch atop the hill on our one early morning walk. Closer now, tragic new vistas opened up before me: a posh hotel, completely collapsed, other houses and businesses lying demolished on either side, several bulldozed clear to the street--and a few under reconstruction.

The airport was as chaotic as ever. Bennet insisted on staying with me right up to the three-person wide jostling line. Then I was on my own. His hovering behavior, followed by a quick departure, increased my anxiety. What was he so worried about? IMC had suddenly abandoned me. I was all alone in Haiti, surrounded by hundreds of people. Strange.

The unruly line was so backed up, slow and chaotic. I began worrying I would miss my plane. Several times the redcaps pushed through with portly well-dressed clients claiming they had an earlier flight. Later I met several of them in the waiting area--waiting for MY flight. Rank—and money—hath its privilege. Again I noticed I was unusually anxious, even a little paranoid for some reason, as if something were going to swoop down and prevent me from leaving Haiti at the last second. I even looked up to make sure no black vultures were circling overhead.

There were no discernable vendors or boutiques inside the airport, as if the place had been stripped bare. At the gate I noticed several people receiving big bags of booze, one prominently marked Barbancourt. *How did they arrange this?* But I wasn't ready for a drink--yet. I felt I had to keep my wits about me.

You will be glad to know I got on the plane. Just as I was settling back into my seat on the first leg of my Air France flight (to Pointe-a-Pitre, Guadaloupe) two disquieting images floated through my mind. The first was little Makenta Paul, my four-year-old in Lilavois. Had she survived the earthquake? I still didn't know. The second was that girl with the beautiful dress and the terrible burn on her face. Could I arrange anything for her in Boston? These troubling loose ends wouldn't let me doze in peace.

It wasn't until I landed in Pointe-a-Pitre and walked into the splendor of a normal airport that I felt more relaxed, my paranoia evaporating, and a tranquility spreading throughout my body. I had been carrying a lot more tension tucked away inside me the whole trip than I realized. Plus a little crescendo right at the end. I was happy to have done it. And happy to be going back home.

A week later, with Fred Stoddard's and the Shriner's help, I was able to arrange a possible admission to their fabulous Boston burn hospital. But when the Petit Goave Beatrice Clinic team tried to find her, sending messages out every way they knew how, she seemed to have disappeared into the mountains. We were never able to find her. Dr. Nick, and later Dr. Peter, tried to no avail. I crossed my mind that maybe she and her mother did hear we were looking for them, but the thought of leaving Haiti for burn surgery in the United States was just too scary.

But what about my little Makenta? I worried about her for weeks until I was finally able to reach her outreach coordinator, Margaret Penicaud. She gave me the happy news that Makenta and all the girls and boys in the orphanage had survived. There had been some minor earthquake damage to their facilities, but the convent and school buildings were mainly intact. While on the phone, almost in the same breath as these good tidings, she asked me if I would sponsor her again.

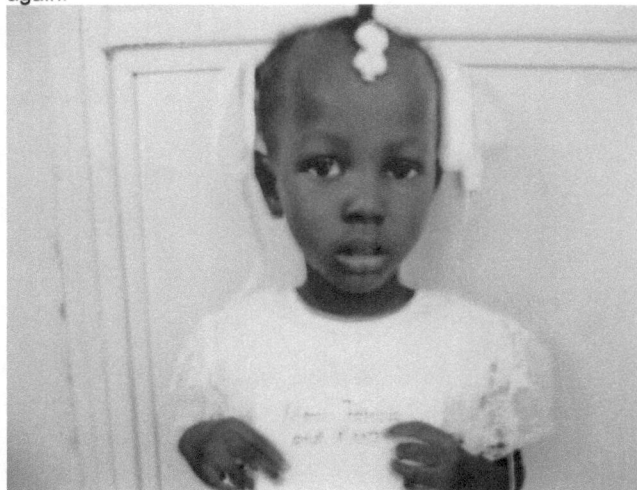

Our sponsored girl, Makenta Paul, now 4, safe and sound, with a 'Thank You' note to us

Relieved and grateful, I immediately said, "Absolutely. I'm so glad she's alive!", the words catching in my throat. As I type these last words, I have tears running down my cheeks again. Haiti is still heavy on my heart, and will be forever. Hopefully her reconstruction won't take that long.

Though I have a very personal, long-standing interest in Haiti, adding particular depth to my experience, anyone doing this kind of volunteer work will find it deeply rewarding, no matter what age and stage you are at in your career. I highly recommend making the adventurous leap of faith to do this kind of thing. It is challenging, but also deeply fulfilling. I hope my clinical narrative inspires you and many others to volunteer. And if you do, that it supports your efforts in your clinical work. For those not in a position to do direct disaster relief work, there is another important opportunity. You can contribute financially to NGO relief organizations, like the International Medical Corps, or Partners-in-Health, two outstanding groups dedicated to training Haitian physicians and nurses to better serve their country.

.

BIOGRAPHY OF KENT RAVENSCROFT, MD

As an undergraduate at Yale, Kent Ravenscroft, a native of St. Louis, began his research into altered states of consciousness. Through his friend, Maya Deren, creator of the definitive documentary on Haitian Voodoo, _The Divine Horseman: The living gods of Haiti,_ he gained access to Haiti's voodoo inner sanctum. Living for a year with a voodoo priest, he attended ceremonies and interviewed peasants possessed by Voodoo gods. His Yale Scholar of the House thesis on Voodoo Possession in Haiti contributed to our modern theory of multiple personality. Going on to Harvard Medical School in 1962 (when his favorite author, Michael Crichton, was there), while becoming a physician, he did sleep and dream research with Ernest Hartmann. During his Harvard residency in psychiatry he collaborated with research hypnotist Martin Orne (later, Sylvia Plath's therapist), continuing research and publishing on voodoo and hypnosis, as well as sleep and dreams.

After graduating in 1966, Dr. Ravenscroft worked with troubled adolescents at NIMH before specializing in child psychiatry. Now Associate Clinical Professor at George Washington and Georgetown Medical Schools, he was Training Director in Child and Adolescent Psychiatry at both institutions, as well as Director of medical and surgical Consultation Liaison at Children's Hospital. In '93-'94 he and his wife took a family sabbatical to France, he commuting to the Tavistock Institute in Hampstead, and she developing her French culinary tour business. A specialist in eating disorders, publishing two edited volumes with Gianna Williams, he collaborated with Sally Bedell Smith on Princess Diana's eating difficulties for her book, Diana: Princess in Search of Herself.

He is known for his clinical and courtroom work in physical and sexual trauma, post-traumatic stress, multiple personality disorder and Munchhausen Syndrome. He was live with Tom Brokaw during the Challenger Disaster and appeared on the Diane Rehm the day of the Oklahoma Bombing, as well as on the Charlie Rose Show. He returned to Haiti for the month of March after the January 2010 earthquake as a volunteer psychiatrist with the International Medical Corps starting a mental health team, serving IMC mobile medical clinics in the Petit-Goave area near Leogane, the epicenter of the quake. He has just completed **Haiti Fare Well**, and this book on this experience there in addition to his first novel, **Body Sharing: the Drug War, the CIA, and Haitian Voodoo;** his wife Patti and he have written **Les Liaisons Delicieuses: Breaking the French Culinary Code**.

Dr. Ravenscroft made his home in Washington D.C for 30 years before moving with his wife Patti half-time to Paris in September 2007. Patti Ravenscroft organizes gastronomic tours to France through her company, Les Liaisons Délicieuses. They now divide their time between Paris and Martha's Vineyard. His son, Christopher, is a chef trained at L'Academie de Cuisine, cooking in Washington; his daughter, Julia, a Savannah College of Art and Design graduate in Fashion works at Louis Vuitton in New York.